Design-Build
SUBSURFACE PROJECTS

Second Edition

Edited by
Gary S. Brierley
David H. Corkum
David J. Hatem

Published by

SME
Society for
Mining, Metallurgy
& Exploration

Society for Mining, Metallurgy, and Exploration, Inc. (SME)
8307 Shaffer Parkway
Littleton, Colorado, USA 80127
(303) 948-4200 / (800) 763-3132
www.smenet.org

SME advances the worldwide mining and minerals community through information exchange and professional development. With members in 71 countries, SME is the world's largest association of mining and minerals professionals.

Library of Congress Cataloging-in-Publication Data

Design-build subsurface projects / edited by Gary S. Brierley, David H. Corkum, David J. Hatem. -- 2nd ed.
   p. cm.
   Includes bibliographical references and index.
ISBN 978-0-87335-321-2
1. Underground construction. 2. Project management. I. Brierley, Gary S. II. Corkum, David H., 1954- III. Hatem, David (David J.)

TA712.D474 2010
624.1'9--dc22
              2010022475

Contents

Contributors

Michael W. Anderson is executive vice president and managing director surety for Willis–North American Construction Practice. He is responsible for serving the team's needs in the production, placement, and management of clients' program capacity, driving innovation in Willis' "Best Practices" for its surety business, and maintaining and enhancing the practice's surety carrier relationships. Anderson began his surety career at Seaboard Surety Company in 1983. Prior to joining Willis in November 2000, he headed Reliance Surety Company's International Department, where he was responsible for management of the company's international underwriting and operations. Anderson's current and past industry activities include national director of Associated General Contractors of America (AGC); member of AGC National Executive Board; and chairman of AGC Service & Supply Council's Executive Committee. He is a frequent speaker on surety industry issues and has published numerous articles and white papers on how contractors can establish and maintain stable surety programs.

Alastair Biggart has worked in the construction industry for 43 years, both as a Contractor and a consultant, specializing in underground works, with particular experience in all methods of underground construction. He has been involved on-site on the Greater Cairo Wastewater Project; the Channel Tunnel; the Storebaelt Tunnel, Denmark; and the Los Angeles Metro North Hollywood extension. Before retiring from full-time work in August 2001, he was tunneling director for Hatch Mott MacDonald. Biggart now works as an independent tunneling consultant.

Brian Brenner is currently with the firm of Fay Spofford & Thorndike and a former supervising structural engineer working with tunnel design and construction on the Central Artery/Tunnel Project in Boston. Brenner has published many papers on tunnel design, automated analysis, engineering education, and other topics. He is chair emeritus of the Editorial Board of the *Boston Society of Civil Engineers Journal* and is editor of the *Journal of Professional Issues in Engineering Education and Practice*.

Gary S. Brierley has more than 40 years of experience with all aspects of subsurface design and construction. He has served as president of the American Underground-Space Association, as chairman of the Underground Technology Research Council, and as a member of the U.S. National Committee on Tunneling Technology. Brierley has published more than 120 articles on the design and construction management of subsurface projects, with a heavy emphasis on geotechnical issues, structural design, and the nontechnical aspects of professional practice, and is a Professional Engineer in 24 states. He is president of Brierley Associates, LLC, a firm that specializes in tunnel design with eight offices and projects throughout the United States.

Charles Button is the deputy chief operating officer for an East Coast water authority and has more than 35 years of experience in design, construction, and operations of water, wastewater, and drainage systems for regional and local public agencies. Working for local and national consulting firms, he has vast experience in design and construction of water, wastewater, and drainage systems in many locations in the United States and internationally. He is the past president and a member of the American Society of Civil Engineers (ASCE) and the Boston Society of Civil Engineers, and a life member of National Society of Professional Engineers and Massachusetts Society of Professional Engineers. He is a registered Professional Engineer in Massachusetts and New Hampshire.

David H. Corkum is a partner at Donovan Hatem LLP. He is a construction attorney familiar with every aspect of the planning, design, and construction of heavy civil and underground construction projects. Corkum's practice is concentrated in the defense of Architects, Engineers, and Construction Managers. He also advises clients on risk management practices and procedures throughout the various stages of a design and construction project. Prior to joining Donovan Hatem in 2001, Corkum spent 23 years as a geologist and construction manager on large-scale and vital energy and infrastructure projects (including Seabrook Station nuclear power plant, the design of the Alaska natural gas pipeline, the site investigation for a high-level nuclear waste storage facility in west Texas, and the construction of the Red Line tunnel extension for MBTA). He has resolved hundreds of disputes between and among Owners, Designers, and Contractors.

Robin B. Dill is a program director in the AECOM Water national wet weather practice group, specializing in tunneling, trenchless technology, and geotechnical engineering. In his 30 years of professional experience, he has developed significant underground-related expertise in conveyance projects involving combined sewer overflow facilities, sewer pipelines, storm drainage pipelines, hydroelectric tailrace tunnels, and electric and gas transmission lines. His tunneling and trenchless project experience includes the use of both rock and soft ground tunneling, pipe jacking, microtunneling, and horizontal directional drilling technologies. In addition to conventional design and construction-related assignments, Dill has experience in Design-Build project delivery and has served in active leadership roles within the Design-Build Institute of America. He has participated as geotechnical leader for major international wastewater facilities using Design-Build delivery, including projects in Bangkok, Egypt, and Singapore. Dill is a registered Professional Engineer in New York, Alabama, Rhode Island, and Vermont.

William W. Edgerton is president of Jacobs Associates. He has more than 30 years of practical, heavy construction experience, including rapid transit systems, dams, power plants, highways, bridges, and marine structures. He has held positions ranging from project engineer to project manager. In the 22 years he has worked for Jacobs Associates, he has provided professional engineering services in the areas of dispute resolution and construction management as well as design. He has managed design teams and served in technical roles for tunnel and deep cut-and-cover projects using both Design-Build and Design-Bid-Build delivery methods. He is a member of the Executive Committee of the Underground Construction Association of SME and a member of the Moles. He is a registered engineer in nine states and Puerto Rico and has published numerous papers on contract risk allocation for construction contracts.

Randall J. Essex is executive vice president and director of tunneling for Hatch Mott MacDonald. He has 33 years of experience involving planning, design, project delivery, construction management, and dispute resolution activities for transit, highway, water, and wastewater tunnels across North America and overseas. Essex has authored more than 40 technical papers and publications, including as principal author of two ASCE publications on Geotechnical Baseline Reports for tunneling and other underground projects. He has presented at numerous conferences, short courses, and universities, is a past board member of the American Underground Construction Association, and is a past chairman of the Underground Technology Research Council.

David Grigg is the national director for professional liability at Willis Construction Practice, where he specializes in working with the design and construction industry in developing effective risk management and risk transfer strategies. Grigg serves on the board of directors of the Civil Engineering Research Foundation of ASCE and on the board of directors of the Building Security Council. Prior to joining Willis, he was a senior vice president and practice leader for the Design Industry Practice, Marsh USA Inc., and a senior vice president and senior program director of DPIC Companies, the specialist professional liability insurance company. Previously, Grigg was vice president of DPIC Management Services Corporation, where he specialized in strategic, business, and marketing planning and management for Architect/Engineer firms. He also has extensive domestic and international experience as vice president of Cadmus Properties, a Design-Build company, and as a principal of Australia Design, which provides technical services to Australian and British companies with operations in the United States.

John A. Harrison, vice president and principal project manager of Parsons Brinckerhoff (PB), has more than 40 years' experience managing large multidisciplinary engineering and construction projects, primarily in the rail transportation field but also in telecommunications and high-energy physics experimentation. As Sound Transit's University Link Project director from November 2006 to July 2009, Harrison directed the final design of the 3.15-mile, entirely underground light-rail transit extension from downtown Seattle to the University of Washington. He is currently deputy program director for the PB-led program management team overseeing the development of a California high-speed train project to connect San Diego, Los Angeles, Sacramento, and San Francisco. This project will involve constructing more than 25 miles of twin-bored tunnels, most likely using Design-Build delivery methods. He is a registered Professional Engineer in eight states.

David J. Hatem is a founding partner of the multi-practice law firm, Donovan Hatem LLP. He leads the firm's Professional Practices Group, which represents engineers, architects and construction management professionals. David has been practising more than 30 years and is nationally recognized for his expertise in representing engineers and architects. He frequently lectures on issues of professional liability for design and construction management professionals, risk management, and design-build procurement issues, and has authored numerous related articles. He edited *Subsurface Conditions: Risk Management for Design and Construction Management Professionals*, *Design-Build Subsurface Projects*, and is co-editor and chapter contributor for *Megaprojects: Challenges and Recommended Practices*, to be published in 2010 by American Council of Engineering Companies. David has served as ACEC/Massachusetts Counsel since 1988 and was the recipient of the 2008 American Council of Engineering Companies of Massachusetts Distinguished Service Award. He was also selected for the sixth consecutive year for inclusion in *The Best Lawyers in America*® 2009 in the fields of Construction Law and Professional Malpractice. He currently teaches Legal Aspects of Engineering at Tufts University.

John Hawley is a senior vice president of Hatch Mott MacDonald and for more than 30 years has managed tunnel design and construction on some of the world's biggest infrastructure projects. Much of his work has been on rail, both inter-city and transit, and he has also led major highway and wastewater projects. His projects include California high-speed train, Silicon Valley Rapid Transit, Los Angeles Metro North Hollywood extension, Hong Kong Strategic Sewage Disposal Scheme, London Heathrow Express, Singapore Utility immersed tunnels, Greater Cairo Wastewater Project, and United Kingdom–France Channel Tunnel.

Gregory G. Henk is vice president with HBG Constructors. He has more than 30 years of experience in the surface transportation industry at the Owner, Consultant Engineer, and construction levels. He has been involved in more than $2 billion of Design-Build projects. Henk served as vice-chairman of Design-Build Institute of America's civil infrastructure committee.

Michael R. Kolloway is the assistant general counsel for operations and risk management for AECOM Technology Corporation since December 2005. He currently sits on the AECOM Office of Risk Management, as well as the AECOM Major Projects Review Committee. Previously, Kolloway was senior vice president and general counsel of Metcalf & Eddy and partner in the Chicago law firm of Rock, Fusco & Garvey, where his practice included defending engineering professionals in a variety of litigation matters and preparing effective documents to protect design/engineering consultants.

Hugh S. Lacy is a partner with Mueser Rutledge Consulting Engineers. He has more than 48 years of heavy underground investigation and design and construction experience, including rapid transit systems, sewer tunnels, utility tunnels, micro tunnels, tunnel shafts, dams, building foundations, and deep basements. His experience, much of which is in the northeastern United States, includes projects countrywide and in Canada, Panama, South America, Taiwan, and the United Arab Emirates. Lacy is a member of the Moles, several ACI committees, the executive board and chair of the George Fox Conference for the Underground Construction Association of SME, and the Association of State Dam Officials, among others. He has published papers on grouting, dam seepage, deep pile foundations, pile load testing, large mat foundations, and tunneling and shafts using artificially frozen ground. He and his firm have been involved in a number of projects that used the Design-Build method of contracting.

Thomas J. Lamb was involved in the design and construction of subsurface projects with Stone & Webster Engineering for 33 years. He was the construction manager for the Massachusetts Water Resources Authority's (MWRA's) Braintree/Weymouth tunnels, shafts, and pumphouse project. Lamb also served as the construction manager for the tunnels and diffusers associated with the MWRA's Boston Harbor project. He was responsible for supervising engineers and geologists in conducting site exploration programs, design, and preparation of specifications for numerous heavy civil and subsurface projects, including the U.S. Department of Energy's high-level nuclear waste repository studies in west Texas, Bradley Lake Hydroelectric Project in Alaska, the Raystown Hydroelectric Project in Pennsylvania, and the Northfield Mountain Pumped Storage Project in Connecticut. Lamb passed away in February 2008.

Cesare (Chet) J. Mitrani, executive vice president of Willis Construction Practice, is responsible for the Texas Region Construction Practice and oversaw the development, administration, and quality assurance of project insurance programs nationwide. He has been involved with more than 300 wrap-up programs during his 39 years in the construction insurance industry. Mitrani has worked on many subsurface projects, including subway construction, mass transit facilities, water and sewer tunnels, and power projects. His background includes safety engineering, underwriting, and risk management. He is a professional member of the American Society of Safety Engineers and has an Associate Risk Management designation.

Thom L. Neff is president of OckhamKonsult, which he founded in 2004 to pursue assignments in risk management and strategic infrastructure management consulting. His professional career has included significant assignments in the planning, research, design, construction, and operation phases of a wide variety of civil and heavy construction projects throughout the United States and overseas. He has also worked on Design-Build and public-private partnership projects that have ranged over transportation, water/wastewater, and oil facilities in Brazil, Saudi Arabia, South Africa, Venezuela, and the United States. On Boston's $14.5 billion Central Artery/Tunnel Project, he was responsible for managing design and construction efforts in the areas of geotechnical, environmental, historic preservation, archeology, and facility deformation monitoring and control. He has worked in the public and private sectors, and held a number of academic positions related to civil, geotechnical, and construction management. He has published extensively in these areas and developed the STEPS Approach to project management, and program issue resolution.

Thomas F. Peyton has spent more than 40 years in the underground industry. He has worked as a Contractor for 25 years building underground tunnels, caverns, and shafts, and 15 years as a Construction Manager of projects such as the MetroWest Water Supply Tunnel project in Boston and Phase 1 of the Second Avenue Subway in New York City. Peyton is the past chairman of the Underground Construction Association of SME.

John Townsend is western regional manager of Hatch Mott MacDonald. In the mid-1980s, he was the resident engineer for the Mass Rapid Transit Corporation in Singapore, on the Design-Build contract for construction of the Singapore River Crossing tunnels between Raffles Place and City Hall Stations. He was also the Designer's site representative as part of Design-Build Contractors' team for the construction of the Medway River immersed road tunnel in southern England. At Hatch Mott MacDonald, Townsend has been involved in Design-Build projects above- and belowground: Whittier tunnel, Alaska; Sound Transit, Central Link, Seattle; and TRAX University and Hospital Lines, Salt Lake City.

Stuart T. Warren, geotechnical manager for Hatch Mott MacDonald, has 35 years' experience in mining and tunneling as a Contractor and consulting engineer, and in managing the geotechnical works on major projects, such as the Channel Tunnel, Storebaelt Tunnel in Denmark, and Los Angeles Metro. He has led site investigations and tunneled in complex geological environments and within a variety of contract forms. Warren has extensive experience with bored tunnel, sequential excavation method, drill and blast tunneling techniques, and the design and supervision of ground treatments for control of groundwater and soil stabilization.

Foreword

David Hatem was kind enough to send me a copy of the first edition of *Design-Build Subsurface Projects* shortly after he and Gary Brierley had completed it. I vividly remember my initial reaction. "What a great topic. Too bad it will be lost on all those geotechnical and tunneling gurus. They'd sooner become yoga instructors than consider working for a contractor on a Design-Build project."

Could I have been more wrong?

It does not take a genius to see that the use of Design-Build on subsurface projects has sort of caught on since this book was first released in 2002. Sort of, indeed. The proliferation of Design-Build has been nothing short of remarkable, particularly when one considers all of the procurement, contracting, and political obstacles that stood in the way of public agencies as recently as five years ago. But, as is well-discussed in this new edition of *Design-Build Subsurface Projects*, the positive attributes and strong performance of Design-Build have become so widely studied and accepted that Owners and their advisors have to seriously consider the system when making their decisions about project delivery. And once they give the process serious attention, they like what they see and find it difficult to use something else.

Almost everyone agrees that Design-Build is a great fit for projects that are schedule-driven. But what about when a project is driven by innovation, design, and other qualitative goals? I had the good fortune of helping the Southern Nevada Water Authority (SNWA) on the Lake Mead Intake No. 3 project, a challenging tunnel project given logistics, geological conditions, and almost 17 bars of hydrostatic pressure. Marc Jenson, SNWA's director of engineering, championed the use of Design-Build not only because of schedule but also because it was the best way to ensure a successful and safe project. Paraphrasing Marc, "We need to have the engineer and contractor figuring this out as an integrated team. And if we reward that team by heavily weighting technical approaches during the evaluation process, we'll have the best chance to get a project that will work. Design-Build gets us there. We can't get there using a Design-Bid-Build approach."

Today it is pretty easy to be a Design-Build cheerleader in the subsurface industry. But let's be clear. Deciding to use Design-Build is one thing; *successfully* using Design-Build is another. And that is why I was honored to write the foreword for the second edition of this book. There is simply no better resource to explain, in a straightforward and comprehensive manner, how to make Design-Build work right on these complicated projects. The reasons I can say this are fairly obvious.

First, the chapter authors are a *who's who* of subsurface experts. Although many of them had Design-Build experience when they wrote their chapters in the first edition, they now have far broader experience: Most of them are currently working on the most visible Design-Build subsurface projects in the world. They are developing best practices on those projects and are sharing it in this book. I find that to be commendable. I did not have the pleasure of knowing many of them until the last few years, when I had the benefit of working with several of these authors on Design-Build projects around the country. Their knowledge base is absolutely superb, and, perhaps more importantly, they bring lots of common sense to the process. This comes through loud and clear in their writings.

Second, as they did with their first edition, the editors have pushed their contributors to deal with the thorny issues of Design-Build. One area that I have found particularly helpful is covered by David Hatem in the area of liability transfer when the owner has prescribed a detailed design solution. The industry sometimes sells Design-Build on the belief that it transfers all liability for design to the Design-Builder under a single point of responsibility theory. In Chapter 3 of the first edition, David Hatem explains the challenges with this approach and suggests a creative approach to balance the practical impact of legal precedent and, for lack of a better word, the *fair* thing to do from a policy perspective. Chapter 3 of this new edition builds on these ideas and gives both Owners and Design-Build teams some great thoughts on how to make the procurement, contracting, and claims avoidance process work better.

My final, big picture thought is pretty simple. I find that this book makes for an interesting read, not only for Design-Build junkies like me who will read anything on Design-Build, but for those who are new to the process. Many readers might be inclined to read the book in bits and pieces. My advice would be to avoid taking any shortcuts. The best way to learn about Design-Build is to have a full understanding of how and why the system works and where the pitfalls can arise in implementing best practices. The authors' years of experience will benefit even the most seasoned of practitioners. I hope that you enjoy their hard work as much as I did.

Michael C. Loulakis, Esq., DBIA
President, Capital Project Strategies, LLC
Reston, Virginia
January 2010

Preface

This book is the second edition of a book of the same name that was published in 2002 by Zeni House Books under the auspices of the American Underground Construction Association (AUA). At that time, very few underground projects had been procured using Design-Build contracting practices, and the primary objective of the 2002 book was to provide some guidance in the use of Design-Build for these highly specialized and somewhat risky undertakings. Needless to say, a lot has happened since 2002:

- The AUA was acquired by SME (Society for Mining, Metallurgy, and Exploration).
- All copies of the original book have been distributed.
- Several major projects involving underground construction have been designed and/or built using Design-Build contracting practices.
- SME recently (2008) published the second edition of the *Recommended Contract Practices for Underground Construction,* which concentrates primarily on Design-Bid-Build contracting practices for underground structures.

The *Recommended Practices* book serves as a template for this Design-Build book, and the two books together are intended to serve as a comprehensive discussion of successful concepts for the planning, designing, and constructing of underground projects. In particular, a list of recommended practices has been added to this edition of the Design-Build book for use by project owners and designers.

The first three chapters of this book provide useful background information about Design-Build procurement. Chapter 1 provides an overview of Design-Build practices. Chapter 2 discusses some of the advantages and disadvantages of using Design-Build contracts. And Chapter 3 discusses risk allocation issues for Design-Build projects, which are significantly different compared to Design-Bid-Build. Following those three chapters are ones about team structure, procurement, agreements, design development, subsurface explorations, geotechnical reports, construction phase issues, and insurance. Chapter 12 provides a compilation of recommendations from Chapters 4 through 11.

There is one issue that the editors of this book would like to emphasize. Much of the literature about Design-Build procurement (almost all of which is related to vertical, aboveground projects) promotes Design-Build and suggests that Design-Build is preferable to Design-Bid-Build with respect to quality, cost, schedule, and risk allocation. The editors of this book do *not* believe that to be the case for underground construction. Both Design-Build and Design-Bid-Build have advantages and disadvantages for underground projects, which need to be carefully weighed and evaluated during the planning period so that a conscious decision about what is best for the project can be made based on all the important variables associated with a project, including third-party impacts and environmental and community concerns. With respect to this issue, the staff at Parsons wrote us about their experiences with Design-Build, and this treatise can be viewed in Appendix A. That information, combined with the thoughts expressed in the various chapters, produce an excellent list of all the concerns that must be addressed in order to produce a successful Design-Build underground project.

CHAPTER 1

Overview of Design-Build Subsurface Projects

Gary S. Brierley, PhD, P.E.
President, Brierley Associates, LLC, Denver, Colo.

David J. Hatem, PC
Partner, Donovan Hatem LLP, Boston, Mass.

Much has been written about the increasing utilization and promise of the Design-Build delivery method. The published literature is robust with articles about Design-Build in the context of vertical (building) and horizontal (civil) construction projects. Numerous surveys demonstrate large increases in both the number of projects and the construction values being procured on a Design-Build basis, and, as a consequence, a number of traditional construction companies and engineering firms are developing or are considering developing Design-Build capabilities. Since 1990, the American Institute of Architects, the Engineers Joint Contract Documents Committee, the Design-Build Institute of America, and the Associated General Contractors of America all have produced standard contract documents for use on Design-Build projects. Clearly, in the last decade, the use of Design-Build has significantly increased in the construction industry, with projections for continued growth.

The primary purpose of this book is to explore and discuss various subjects regarding the selection and implementation of Design-Build procurement practices for the design and construction of *subsurface projects*, defined as projects involving substantial amounts of underground work such as tunnels, highways, dams, and deep foundations. While some of the general Design-Build literature certainly is valuable, the application of Design-Build to subsurface projects poses a number of special challenges and opportunities that merit a more focused consideration. The various chapters of this book strive to achieve that objective.

A secondary purpose of this book is to provide Owners, Engineers, Construction Managers, Contractors, and others involved in the design and construction of subsurface projects with a conscientious survey and discussion of the relevant factors to consider in selecting and successfully implementing Design-Build for subsurface projects. Although Owners are the principal decision-makers in selecting Design-Build as their delivery method, Engineers, Construction Managers, Contractors, and many others serve important consulting and advisory roles to Owners in the project-specific implementation of Design-Build. This book is not intended to be an advocacy piece for the use of Design-Build for subsurface projects but rather to provide an objective, focused, and balanced discussion of the subject matter.

Many written articles compare the use of Design-Bid-Build to Design-Build, with the implied assumption that Design-Build is somehow the preferred alternative. Almost always, these articles identify the following advantages of Design-Build:

- **Single-point responsibility:** The Owner contracts with only one entity, the Design-Builder, which is singularly responsible to the Owner for design and construction of the project.

- **Design-Builder control:** The Design-Builder has more control over the work and increased opportunity to participate in the design development process, constructability reviews, and innovation in reducing cost and schedule.

- **Increased incentives/cooperation:** The Designer and the Builder have increased opportunity to collaborate and minimize disagreements about inappropriate, conservative, or unnecessary design requirements.

Clearly, Design-Build is appropriate under many circumstances and, in fact, results in many of these advantages. However, most of the successful Design-Build case histories in the United States are associated with aboveground structures designed and constructed for private clients by well-integrated and experienced teams of Constructors and Engineers. Most often, these teams become involved with the delivery of a single or standard type of facility, such as apartment buildings, manufacturing facilities, or power plants that are constructed on land owned by the client with minimal impact on either third parties or the environment. This approach to Design-Build greatly reduces the upfront costs for business development and conceptual design, may significantly reduce the schedule, and provides the Owner with a high degree of confidence that it will obtain a well-designed and well-constructed finished facility with specific long-term performance characteristics.

Historically, almost all construction was performed on the basis of the Design-Build method with an Owner retaining the services of a master builder who coordinated all aspects of design and construction. Beginning around 1900, however, many Owners began to separately retain an Engineer to provide only design services and a Contractor to construct the project. Far from being a flawed system of project delivery, the Design-Bid-Build method also offers many advantages, such as

- **Owner control:** The Owner retains control of the design process and can respond more fully to the financial, political, or community concerns associated with the project.

- **Focused responsibility:** The Owner works with an Engineer to determine exactly what needs to be constructed for both short- and long-term performance criteria and then with a Contractor to actually build what is required and specified in the contract documents.

- **Final and complete design:** The Contractor receives a contract document describing all the work that must be accomplished during construction and allowing Contractor to price the final design.

- **Construction management:** The Owner retains more control during construction, especially with respect to the quality of the completed project work. In addition, the presence of the Owner's Engineering Consultant during construction, while adding some tension in the process, provides a degree of protection and representation for the Owner, as well as checks and balances in the execution of the design in accordance with the contract documents.

In reality, these advantages of Design-Bid-Build serve many Owners well. All projects begin as an idea and end as a finished facility. At some point, an Owner, either public or private, decides that it must build in order to accomplish an economic objective or fulfill a public need. Is it possible to determine the best delivery methods to convert the idea for a project into a finished facility? Can the Owner take advantage of the best that Design-Build has to offer without sacrificing too much control over the highly complex and risky planning scenarios that are required for subsurface projects?

Subsurface projects carry significant risks for Owners, Engineers, and Contractors. The risk of disappointed contractual and commercial expectations of project participants is substantial on any subsurface project, regardless of whether the delivery method is Design-Bid-Build or Design-Build. The essential components for the successful implementation and delivery of any subsurface project for both Design-Bid-Build and Design-Build contracting practices are

- Timely identification of risk potential,
- Clarity and adequacy in communication to project participants of relevant information regarding that risk potential, and
- Fairness and clarity in the allocation of that risk among those participants.

Those involved and experienced in the design and construction of subsurface projects utilizing the traditional Design-Bid-Build method have, especially during the previous 30 years, learned some valuable lessons about risk identification and allocation. During that time, a series of *improved contracting practices* for those involved on subsurface projects has evolved. Although Design-Build certainly produces good reasons why some specialized risk allocation principles and practices need to be developed and adapted for Design-Build subsurface projects, the *lessons learned* from improved contracting practices developed in connection with Design-Bid-Build subsurface projects have relevance and value in Design-Build, and it would be a serious mistake and regression for the subsurface design and construction industry (and, most especially, for the Owners whom it serves) to ignore and disregard them in implementing the Design-Build approach. Fairness and clarity in risk allocation are critically important to the successful implementation of Design-Build on subsurface projects. Some Owners believe that the risks of differing site conditions and defective design should be *entirely* allocated to the Design-Builder, even though the former may undertake and control some or all of the subsurface investigations and, to a substantial degree, the development of design. Are those beliefs and objectives appropriate reasons for Owners to select Design-Build on a subsurface project? In the opinion of these authors, Design-Build neither mandates nor depends upon the absolute or total risk transfer to the Design-Builder, nor does such an approach make sense. History repeats itself, and the absolute/total risk transfer approach will predictably lead to the same failures, disputes, and other negative consequences that have resulted from similar approaches in the Design-Bid-Build method. The chapters of this book address the challenge of how to translate and adapt the improved contracting practices and lessons learned in subsurface projects from the Design-Bid-Build experience to the Design-Build context.

This book examines many specific subjects regarding the use of Design-Build on subsurface projects. Is Design-Build right for a particular project? Why and under what circumstances should an Owner consider Design-Build appropriate for use on a subsurface project? What are the appropriate factors to consider in making this decision? What are the wrong reasons for choosing Design-Build? Is the Owner prepared and willing to relinquish the traditional degree of control over the design development and finalization process and to delegate significant design development discretion and control to the Design-Builder? When should an Owner retain an independent Engineer in Design-Build, for what scope of services, and during what phases of the subsurface project? What should be the role of the Owner's Engineering Consultant in advising the Owner as to whether the Design-Build method should be used on a subsurface project? Is the Owner willing to share risk with the Design-Builder? What has been the experience to date with the use of Design-Build on subsurface projects? How can we continue to learn about and share positive and negative experiences regarding the use of Design-Build on subsurface projects? What is the future of Design-Build on subsurface projects? How much subsurface site investigation should be undertaken? By whom? How much subsurface investigation information and opinion should be shared with the Design-Builder? Should that information be disclaimed as to accuracy? What geotechnical and related reports should be prepared, by whom, and for what purposes? Should any of those reports become part of the Design-Build agreement? Should the Design-Builder have the right to rely upon subsurface information, opinions, or reports furnished by the Owner? How should risk be allocated for subsurface conditions? Should the Design-Build agreement contain a differing site conditions clause? Should the Owner prepare and furnish preliminary design and/or performance specifications and, if so, developed to what extent or degree? Who should be responsible for the final design? Should design responsibility be bifurcated between the Owner-furnished preliminary design and the Design-Builder's final design?

Other issues must be acknowledged in considering the transformation from Design-Bid-Build to Design-Build in the subsurface project context. One aspect of this transformation is *contractual* and includes the definition of different professional and contractual relationships among project participants; differing scope of services, work, responsibilities, and risk assumptions; and the assumption of different levels of control (typically less for the Owner and more for the Design-Builder) commensurate with the differing allocation of risks. Another equally important aspect of this transformation is *behavioral*. As the roles, risks, and responsibilities of project participants in Design-Build change from those traditionally undertaken in Design-Bid-Build, so must the participants adjust or modify their behavior to accommodate and align with the new roles, risks, and responsibilities. For example, an Engineer who is a subconsultant to the Design-Builder does not owe its allegiance or professional accountability to the Owner, and the Owner must understand that in Design-Build it must relinquish significant control over design development in exchange for its ability to transfer more risk to the Design-Builder. In addition, the Design-Builder has affirmative obligations to produce an adequate final design and may not simply construct based on a preliminary Owner-furnished design.

Putting the right contract in place does not necessarily guarantee that the *behavior* of project participants will correspondingly change to align with the *contractual* transition. However, both the *contractual* and *behavioral* transitions must occur in order for Design-Build to be successful. These questions and issues, and the challenges they pose, are important for the evaluation and successful implementation of Design-Build subsurface projects. The various chapters of this book explore these questions and issues and the corresponding challenges.

The evaluation and utilization of Design-Build is increasing, and this trend is expected to continue. No definitive answers to the many challenging questions and issues raised and discussed in this book may honestly or definitively be offered at this point. The authors modestly hope that this book will add significantly to the knowledge base regarding use of Design-Build on subsurface projects. We invite and welcome your comments, criticisms, sharing of experiences, and other interactive dialogue. For these reasons, we envision and consider this book to be a living document.

CHAPTER 2

Why Design-Build?

Thomas Lamb, P.E.
Deceased, formerly with Stone & Webster, a Shaw Group Company, Stoughton, Mass.

Gregory G. Henk, P.E.
Vice President, HBG Constructors, Rancho Santa Margarita, Calif.

INTRODUCTION

The Design-Build delivery system is not new. The basic concept almost surely predates the Design-Bid-Build system widely used today on public works projects, and the Design-Build system has a long and successful history in the private sector. Indeed, anyone who has contracted for a new home or substantial renovation has probably experienced both the rewards and the frustrations of the Design-Build delivery system. In the right circumstances, the Design-Build approach offers significant advantages in terms of schedule and accountability. Because of these advantages, the method has enjoyed increased popularity in recent years, especially among public Owners seeking relief from the cost overruns, schedule constraints, claims, litigation, and seemingly endless finger pointing among the Contractor, Owner, and Engineer over which is to blame. Studies by the Construction Industry Institute and the Design-Build Institute of America, among others, demonstrate that the Design-Build method, properly used, results in less cost growth, less schedule growth, and comparable quality when compared to the traditional delivery systems used in public construction.

On the other hand, Design-Build is not a panacea. Owners give up a significant amount of control over the details of the final design when the Design-Build system is selected. In addition, some of the checks and balances of the Design-Bid-Build approach—resulting from the independence of the Contractor and Engineer—are lost in Design-Build. Furthermore, changes or constraints introduced by the Owner or by third parties can quickly negate any advantages of Design-Build and result in the same delays, cost overruns, and lawsuits that plague the traditional delivery systems. A successful Design-Build project requires clear definition, control of outside influences, and an Owner that is willing to step back and let the process work.

ADVANTAGES OF DESIGN-BUILD

The advantages of the Design-Build system most frequently cited are

- Shorter schedule,
- Lower cost growth (although not necessarily lower cost), and
- Single-point accountability.

Certain schedule advantages over Design-Bid-Build are inherent in the Design-Build system. Perhaps the most obvious results from eliminating the bidding cycle. For a complex project, the advertise-bid-award cycle can easily occupy four to six months—and much longer if rebids or protests are involved. With traditional methods, the project is essentially on hold during the bid cycle (although permitting and similar activities are frequently ongoing), simply because the design must be 100% complete prior to bidding and cannot subsequently be changed, except by addendum, which can lengthen the bid cycle. Construction cannot begin until bids have been evaluated and the contract has been awarded. Eliminating the bid cycle eliminates a dead spot in the schedule and can significantly reduce overall project duration. However, some of the time saved may be consumed in the selection of a Design-Build team, because selection of the team is typically more complex and time-consuming than selection of an Engineer alone.

Probably the most significant advantage of Design-Build is the enhanced opportunity to *fast-track* a project. In this approach to design and construction, the design activities are phased such that procurement of long lead-time equipment and actual field construction may begin well before the design is 100% complete. For example, a power plant's boiler and turbine may be selected and ordered early in the project—at the very start of design. Indeed, it is a great advantage to do this, because the auxiliary systems, as well as the foundations and structural supports, may then be designed around the specific equipment to be purchased. Once the size, configuration, and weight of the major equipment are known, economical foundations and supports may be designed. Pipe sizes and configurations can be optimized. Auxiliary equipment, such as pumps and fans, may be selected for optimal performance, and the buildings may be sized and configured to provide clear access for construction and maintenance, yet minimize wasted space. The same principle applies to many other types of projects.

Site preparation and erection of construction facilities, access roads, and utilities can almost always begin before the design of a facility is complete. Demolition of existing facilities, installation of cofferdams, and subsurface improvements are other activities that can proceed well in advance of a complete design. Earthwork and foundation preparation have built-in delays where surcharging and consolidation are employed. Many projects utilizing the traditional Design-Bid-Build approach recognize this and shorten their overall schedules by allowing separate contracts for these activities in advance of the main construction contract. Coordinating the schedules of the various Contractors can be a major headache with this approach, particularly if one of the early contracts is delayed by a subsurface surprise. The Design-Build method offers considerably more flexibility in terms of workarounds and possibilities for mitigation than does the traditional method.

Equally important, Design-Build does not require that the start of construction be held up while all design issues are resolved. Design details, especially those related to finish work, often have little or no impact on other parts of the work but can take considerable time to resolve. The traditional Design-Bid-Build method requires that all, or nearly all, of these details be resolved prior to the release of the documents for bids. With Design-Build, construction can begin before all the details are worked out. It is important, however, not to leave too many loose ends when construction begins, because unresolved issues can cause future delays.

In the most extreme form of fast-track construction, design drawings and specifications are issued just in time to support construction, and the two activities proceed in parallel. Time savings can be rather dramatic. However, the fast-track process is difficult to manage. Delayed design drawings due to design problems or third-party issues can cause delays in construction and substantial additional costs. The cost of such delays is much greater when construction is already underway, as daily costs on a major project can easily exceed $100,000 to $200,000. The process is easier to manage under the Design-Build method, because the Engineer and the Contractor are the same entity, and, in theory, will take the necessary steps to mitigate the delay.

Design-Build can also, in principle, reduce construction time in more subtle ways. Selection and approval of equipment and materials can be done in a much more informal and efficient manner than with the traditional submittal process. Note, however, that making this process more efficient also reduces the Owner's input to and control over the process. Similarly, interpretations of the plans and specifications can be much less formal and will certainly be less controversial when the Engineer and Contractor are one. Conflicts and interference should be identifiable at least as early as in traditional methods and can be resolved more quickly. Early participation by the Contractor in the design development process allows for a thorough constructability review and selection of a design that can be constructed in minimum time.

From the Owner's perspective, one highly significant advantage of Design-Build is the single point of contact and responsibility. In Design-Bid-Build, separate Engineers and Contractors, both working independently for the Owner, often perceive themselves as adversaries. The Engineer worries about the unscrupulous Contractor cutting corners, taking shortcuts, substituting inferior equipment or materials, or misinterpreting the specifications, resulting in an unsatisfactory final product or extra costs to achieve the product the Engineer had in mind. Engineers are also well aware that Owners may look to them or their insurance carriers to correct the problems or contribute to the costs related to defective specifications. One result of this is specifications that are too often conservative, restrictive, and behind the state of the art, and general conditions that attempt to pass all the risks to the Contractor. Engineers complain bitterly about Contractors that perform to the minimum standard and file claims over trivial omissions and ambiguities in the specifications.

Contractors perceive Engineers as stodgy keepers of the status quo, out to stifle innovation and prevent the Contractor from being creative and thereby increasing profits. Contractors see contracts as one-sided and specifications as often ambiguous or even misleading, with the Engineer always insisting on the most expensive interpretation. Contract language favored by Owners reinforces this view. Low margins and intense competition prevent Contractors from carrying much contingency in their bids, so the Project Manager must watch every penny. The practice of accepting the lowest bid actually encourages the Contractor to cut corners in order to win the job. The result is a Contractor that does the minimum required by the specifications and that tries to innovate to the best of its ability within the often restrictive confines of the contract, while complaining that the Engineer is unrealistic or "doesn't understand construction."

This adversarial relationship leads to many disputes and disagreements between the Contractor and the Engineer. These disputes divert the participants' attention from the main task of completing the project and can burden the Owner with mediating them. Many ideas have been advanced to reduce the adversarial relationship, including dispute review boards and partnering. It is the authors' experiences that these methods work well and reduce claims if the Contractor is able to turn a profit. However, if the Contractor is losing a substantial amount of money, regardless of fault, the best these methods can accomplish is to keep the project moving and the arguments civil. If necessary changes in the work exceed the Owner's budget, the same adversarial relationship develops.

Design-Build, in principle, avoids this adversarial relationship between the Engineer and the Contractor, because, as the same entity, it should be less difficult for the Owner to manage. Although the Owner avoids the need to involve itself in disputes between the Engineer and the

Contractor, the Owner loses the independent consultant that, in the traditional method, guards the Owner's interests. With Design-Build, the Owner must look out for its own interests or hire an intermediary to do so. However, Owner oversight can be at a higher level, assuming a proper performance-based specification for the final product has been prepared and the Owner need not get involved in the details. There is also a great advantage to having a single point of contact for all Owner questions and concerns. Owner issues cannot be avoided by laying responsibility on the other party, and the Design-Builder has an incentive to arrive quickly at the most cost-effective solution.

Quality issues must be approached somewhat differently in the Design-Build method. With the Design-Bid-Build method, either the Contractor is required to have a quality program, with oversight and checks by the Owner, or the Owner's Engineering Consultant provides inspection and acceptance testing. In both cases, the objective is to document conformity with the minimum requirements of the specifications and applicable codes. The Engineer is required to address deviations from the specifications and codes, and provide a resolution consistent with the performance requirements of the project. In the Design-Build method, there has been concern that quality issues are subordinated to cost and schedule issues, especially in Contractor-led teams, because the Engineer is no longer independent. To the authors' knowledge, this has not happened to any significant degree.

Indeed, improved quality is cited as an advantage of the Design-Build method. On one level, Design-Build allows quality to be factored into the selection process and even assigned a value. Perhaps more importantly, the selection criteria for a Design-Build contract typically include reputation and past experience, not just price. Thus, both the Contractor and the Engineer, especially the latter, have a definite incentive to produce a quality product. Finally, the responsibilities of the Design-Build team under warranties and guarantees are considerably broader and longer lasting than the Contractor's traditional responsibilities, because the warranty of the design, typically provided by the Owner, is provided by the Design-Builder.

Studies have shown that Design-Build leads to lower cost growth than Design-Bid-Build. This is doubtless due in part to the Contractor's greater freedom to innovate, but probably also to the changed playing field and the different allocation of risk. With traditional delivery methods, Contractors are accused of bidding low with the expectation of making up any shortfalls through extras and claims. While it is not known how often this happens, if at all, it is a common perception among Owners and Engineers. Such a strategy is not feasible with Design-Build, because the prospects for claims and extras for anything but differing site conditions are limited. The Owner is largely shielded from the cost increases associated with design changes. This, in itself, is a distinct advantage for the Owner and greatly reduces the potential for cost increases.

There are also less obvious savings to the Owner in the form of reduced management costs. Because the Owner's degree of involvement in project details and decision-making responsibilities are reduced, the Owner needs less staff to manage the project. Savings can be found in the areas of project management, quality assurance/quality control, change order and claim processing, and litigation. In some cases, management costs can be reduced by a factor of five or more. However, the amount of the reduction depends on how much the Owner is willing to rely on the Contractor for information, and this decision has a definite political aspect.

On the other hand, the Design-Build method generally requires the Contractor to take on more risk than with Design-Bid-Build, and the risk is less well-defined and, consequently, more difficult to manage. In the traditional system, the Contractor's risks are defined in the contract, and the Owner bears all other risks. With Design-Build, it is more common to define the Owner's risks and leave all others to the Design-Builder. Depending upon how the contract is structured and how risk is allocated between the parties, the Contractor may be asked to bear the risks of unforeseen design problems, existing conditions within rather wide limits, regulatory uncertainties, changing economic conditions, labor market uncertainties, and many others. Some of these risks can be managed via insurance, foresight, and careful management. Others are largely beyond the Design-Builder's control and must be dealt with in other ways, such as contingencies. It is unrealistic to think that a Contractor will take on additional risk without additional compensation, so the increased risks should lead to higher costs, all other things being equal. Whether the efficiencies associated with Design-Build outweigh the costs associated with the increased risks, thereby resulting in a less costly project overall, depends on the details of each individual project.

DISADVANTAGES OF DESIGN-BUILD

In some circumstances, Design-Build is not a particularly effective method. Using it for such projects will probably lead to disappointment on the part of all participants. Such circumstances include

- Owners who want control over all the details of the final design,
- Uncertain regulatory environments,
- Significant third-party constraints,
- Unresolved environmental issues,
- Unknown and potentially devastating (to costs or schedules) site conditions, or
- Unrealistic expectations for risk sharing by one or more of the parties.

Design-Build works best when the Contractor is given maximum freedom to innovate. This means the final product should be described in performance terms wherever possible, and acceptance should rely largely on performance tests and guarantees. This is not to say that inspection and testing during construction should be ignored; only that how the various parts fit together should be left to the reasonable discretion of the Design-Builder.

Some Owners have good reasons for not wanting to leave decisions as to the type, size, and number of pieces of equipment up to the Design-Builder. Consider an Owner that is undertaking a hydroelectric plant. Head and flow information can be used together with costs and efficiencies of the turbines, generators, and associated equipment and the value of the power produced to optimize the number and size of the turbines. However, the Owner may have reasons, such as ease of maintenance or the presence of other units in its system, for wanting to standardize and use a particular make and size of unit. Similarly, a sewer authority may want to avoid the problems and expense associated with stocking spare parts for a variety of different pumps and standardize on a particular manufacturer and a limited range of sizes, even if the result is not optimal. Some Owners may insist on certain standard details, which have served them well in the past. While Design-Build can accommodate some specific Owner requirements, the differences between Design-Build and Design-Bid-Build soon become blurred, and much of the advantage of Design-Build is lost. Some Owners simply want to select all the details of their projects themselves and be deeply involved in the construction process. Design-Build will leave such Owners frustrated and their Design-Builders regretting they signed the contract. In addition, an uncertain or changing regulatory environment can quickly lead to massive schedule delays and cost overruns. No Design-Builder, and few Owners, can take such risks, and a different type of contract is needed in such cases.

Many public Owners find it necessary to negotiate memoranda of understanding (MOU) or similar agreements with local communities to win acceptance for their projects or to mitigate their impact. These agreements typically involve restrictions, sometimes significant ones, on design or construction of the facility or both. Such restrictions might include height limits, landscaping requirements, and restrictions on work hours, haul routes, parking, noise, and access to the site. Design-Build can accommodate such third-party agreements if certain conditions are met. These include executing the MOUs *before* the Design-Build contract is finalized, including only reasonable restrictions (noise limitations are sometimes set at levels impossible to achieve on a consistent basis), setting up and specifying a complaint reporting and enforcement mechanism, and, perhaps most importantly, making sure the community truly understands the impact that will inevitably occur. Unrealistic expectations among people living adjacent to the project cause problems for all involved.

These requirements sound simple, but they can be difficult to achieve in practice. Since the design of the project is at a preliminary stage when the Design-Build contract is executed, it can be difficult for community leaders, who are usually not design or construction professionals, to envision the final product. This makes it difficult to agree on restrictions that address the community's goals yet allow the Design-Builder sufficient latitude to design a functional and efficient project. Construction restrictions involve difficult negotiations, and the negotiators must be careful not to agree to something that sounds reasonable but which can have a devastating effect on construction efficiency.

Other third-party concerns, such as abutters who vehemently oppose the project, organized community opposition, or lack of governmental support, may raise the risks of substantial delays or cost increases to such a level that Design-Build is impractical. Contractors are not in a position to deal with this type of opposition, which is best handled by the Owner. Similarly, if the site contains known or potential contamination, regulatory agencies may step in and dictate what can and cannot be done. In these cases, the Owner cannot avoid the risk, which statute assigns to the Owner.

Allocation of risk among the parties to a Design-Build contract is discussed at length elsewhere in this book (see Chapter 3) and will be mentioned only briefly here. Owners cannot expect to transfer all the risks of existing site conditions to a Contractor. In some venues, the Owner's responsibility for differing site conditions is statutory. In others, and for subsurface construction in particular, it is customary to limit the Contractor's exposure through some contractual mechanism. If conditions at the site (including subsurface conditions, utilities, pollution, easements, deed restrictions, and conditions of existing structures) are not well understood, and if there is a reasonable risk of encountering unknown conditions, which, if present, will substantially increase the cost to complete the project, the Contractor will be (or should be) reluctant to assume the risk. In such cases, the Owner must undertake a program to better define the conditions, or the contract must contain provisions for renegotiation if unforeseen conditions are discovered. Some highway projects have successfully used a double notice to proceed (NTP) process, in which a limited NTP permits mobilization and commencement of design, but funds are restricted until issues are resolved and a full NTP is issued. Where conditions cannot be reasonably well defined, Design-Build may not be the most appropriate delivery system.

DESIGN-BUILD FOR HEAVY CIVIL AND SUBSURFACE CONSTRUCTION

Design-Build has been little used in the United States for dams, tunnels, caverns, canals, pipelines, and other heavy civil construction, with the notable exceptions of mining and development of storage caverns in salt. The reasons for this are not completely clear, because in few other areas of construction are the design and construction methods so intimately related. In part, the relative unpopularity of the method probably stems from the fact that many of these projects involve public Owners, which either by regulation or tradition have used other project delivery methods (as opposed to mining, where much of the design work is done in-house). It is also likely that the difficulty of defining subsurface conditions prior to design and the resulting uncertainty in cost has discouraged Owners, particularly public Owners, from pursuing Design-Build. Also doubtlessly contributing to the problem are: difficulties in defining the best value when the design is only conceptual, ensuring a reasonable level of competition, maintaining an adequate degree of independent Owner representation in the design and construction process, and justifying selection of a Design-Builder (which may not be the lowest bidder) to skeptical boards of directors and public officials. However, with increased interest from public Owners in Design-Build, it is appropriate to examine how the method can be applied to heavy civil and subsurface construction.

As noted, in subsurface construction, design and construction methods are closely related. The very process of excavating a subsurface tunnel or chamber drastically changes both the state of stress around the opening so formed and the properties of the materials adjacent to the opening. The extent of the changes depends not only on the size and shape of the opening but also upon the method of excavation, the type of support, and the timing of support installation. Tunnel Contractors tend to be innovators (some would say they have to be to survive), and traditional Design-Bid-Build contract documents generally contain alternate designs or provisions for substituting Contractor designs, particularly in the areas of tunnel support, excavated diameter, and lining systems. Despite this flexibility, Contractors frequently propose something that the Engineer did not contemplate or address in the contract documents. Thus, it would seem that subsurface construction in particular would be a prime candidate for the Design-Build method, if only to foster creativity and innovation. In addition, the inefficiencies and compromises associated with accommodating a variety of means and methods, as well as Contractors with different levels of skill, would be avoided.

Dams are another type of project that could benefit from the Design-Build method. In a dam project, the key to success is the construction sequence, which must accommodate diversion facilities (often incorporated in some form into the final project), seasonal variations in flow, weather, equipment fabrication schedules, and often remote locations lacking infrastructure. While the Engineer in the traditional Design-Bid-Build method can (and must) make fairly accurate assumptions about the construction sequence, the Contractor is the one who actually determines it. Departures from the Engineer's assumptions can require revision of the design, with potential consequences on the project cost and schedule.

The benefits of Design-Build for heavy civil and subsurface projects are similar to those associated with other projects. Reduced schedule is the primary potential benefit, and the potential savings are even greater than on most other projects. Subsurface projects require extensive, costly, and time-consuming site exploration programs, and the lead time for equipment such as a tunnel boring machine (TBM) or turbine generator can be a year or more. Design-Build allows the major equipment to be selected early, and procurement and design can proceed simultaneously. Site work can also proceed while site explorations are ongoing, *provided* enough information is available to confidently site all the major facilities. Relocating a structure after construction has started negates any cost or schedule savings. Reduced cost escalation and single-point responsibility apply to subsurface projects also.

Design-Build allows the design to be tailored to the Contractor's available equipment, materials, and expertise, and allows the entire Design-Build team's past experiences with particular construction methods or with local conditions to be factored into the design. With Design-Build, it is not necessary to target the design at the lowest common level of Contractor skill; full advantage can be taken of the team's particular expertise. If, for example, the Contractor owns or can obtain a used TBM, it is a simple matter to design the tunnel support and lining systems, as well as water control provisions, to take advantage of the particular features of that machine and avoid the incompatibilities that arise so often on Design-Bid-Build projects. Similarly, the personnel available to the Contractor may be more comfortable or more efficient with certain construction methods; as these can be specified, less-effective methods (which might be the best methods for a different Contractor) need not be considered.

SUMMARY

A successful Design-Build subsurface project needs to satisfy certain requirements:

- The project must be defined in enough detail at the conceptual stage to solicit competitive proposals. Design criteria and performance requirements must be clear enough so that all proposers have the same understanding of the final completed project requirements. It is unlikely that Owners, especially public Owners, will not require competition, whether the competition is qualification-based or cost-based.

- Risks of unforeseen conditions, especially subsurface conditions, must be dealt with in a realistic and fair manner. Differing site condition claims are common on subsurface projects, even those with extensive exploration programs. With Design-Build, much less is known at the proposal stage, and the chances that things will not be as expected are that much greater.

- The Owner must be flexible as to how performance requirements are satisfied and must be willing to step back and relinquish significant control and influence over the design and construction process. An Owner that wants to be involved in the selection of equipment and architectural finishes is better off using the Design-Bid-Build method.

- Clear-cut acceptance criteria must be established. This difficult but important step is essential for both the Owner and the Contractor, so that successful completion of the project is clearly defined and unambiguous.

- Regulatory requirements must be clear. An environmental impact assessment is required for every project of any significance, and the predicted impacts are related to the design. It is extremely important to not over-specify the design during the permitting process and lose the flexibility otherwise associated with the Design-Build method. This can be difficult, because regulatory agencies usually want specific commitments in the permits. If the Design-Builder is given responsibility for obtaining permits, a significant amount of additional uncertainty is introduced into the process.

- A fair and equitable payment methodology must be established that recognizes the upfront costs associated with design and subsurface exploration, yet provides clear and unambiguous mechanisms for measuring progress.

- Third-party constraints must be defined and controllable. Agreements with communities must be specific and realistic. Also, politically powerful community groups and governmental agencies must be restrained from demanding more or costly concessions at a later stage. This is a difficult task, because not all impacts can necessarily be foreseen, and both a politically savvy Owner and a flexible Design-Builder are needed to address new issues.

- Funding must be available to support the project's projected cash flow. Any cost or schedule benefits will quickly evaporate if the pace of the project is dictated by available funding.

A subsurface project that can satisfy these preceding requirements is a prime candidate for the Design-Build method. While considerable effort and certain institutional changes will be necessary to implement the Design-Build method on subsurface projects, the potential rewards are significant. Considerable political pressure is building for wider use of Design-Build in the transportation area and will likely extend to other infrastructure projects, including subsurface projects. The potential for reduced schedules, less uncertainty in final cost, and clearer lines of responsibility will drive public Owners to try Design-Build.

BIBLIOGRAPHY

American Society of Civil Engineers. 1998. *Journal of Construction Engineering and Management*, November–December.

Construction Industry Institute. 1997. "Project Delivery Systems—Which One?" Austin, TX: CII.

Florida Department of Transportation. 1994. Transportation Research Record 1351. Tallahassee, FL: FDT.

U.S. Federal Highway Administration. 1996. "Design-Build—FHWA's Role in the Design-Build Program Under Special Experimental Report Number 14 (SEP 14), 10/96." FHWA Report. As reported in Government Accountability Office Report to Congress, March 6, 1997.

CHAPTER 3

Risk Allocation for Subsurface Conditions and Defective Design

David J. Hatem, PC
Donovan Hatem LLP, Boston, Mass.

INTRODUCTION

Many risk allocation issues are associated with the Design-Build method generally and, more particularly, on subsurface projects. This chapter will address the two most important of those issues: (1) risk allocation for subsurface conditions; and (2) risk allocation for defective design. As an overview comment, the legal implications of risk allocation decisions in the Design-Bid-Build method are better understood than in Design-Build and, hence, more predictable, given the well-established legal precedent and documented experience available. At this point, there have been relatively few legal precedents dealing with these risk allocation issues in the Design-Build context, and it is expected that the continued growth and use of alternative dispute resolution mechanisms, including dispute review boards, especially in subsurface projects, will further slow the development of legal precedent (Hinkle 2001). Despite the paucity of legal precedent, there is a well-developed and valuable body of experience, "lessons learned," legal precedent, and principles that have emerged from the Design-Bid-Build context that are relevant and may be adapted to the Design-Build context.

Two overriding principles apply to the discussion of these two risk allocation subjects:

- The successful implementation of Design-Build depends upon the timely identification and fair allocation of risk among project participants.

- Design-Build does not require or depend upon the transfer of *all* risk from the Owner to the Design-Builder.

Hence, what can we learn from the Design-Bid-Build experience on subsurface projects that will assist in achieving these two overriding principles in Design-Build subsurface projects? This chapter will explore how some of the lessons learned from Design-Bid-Build may be adapted and applied in the context of Design-Build subsurface projects. Much has been written about the frustrations and limitations that result from the inability in Design-Bid-Build to integrate and coordinate, in a timely manner, the respective roles of the geotechnical and design engineer(s) with those of the Contractor, especially in the specific context of projects with a significant subsurface construction component (subsurface projects). These perceived shortcomings and disadvantages include

- The inability to obtain timely and qualified input from the Contractor during design development regarding constructability considerations and the scope and definition of preconstruction subsurface investigation, preparation of reports, and development of information, all of which would be beneficial to the Contractor in the planning and execution of the work, including Contractor's contemplations regarding the means, methods, techniques, sequences and procedures of construction ("construction means and methods"), and selection of equipment; and

- The Engineer's lack of knowledge during design of the specific construction means and methods, and equipment choices and decisions that will be made and implemented by the Contractor in the performance of the work, and the consequent inability of the Engineer to anticipate or predict whether those Contractor choices will influence or impact design assumptions and/or the behavior of site conditions during construction to a degree requiring reevaluation of design decisions.

The ability to overcome these perceived shortcomings and disadvantages is especially challenging on publicly procured projects in which, typically, construction contracts are awarded to the lowest bidder based upon complete design documents. In this context, the opportunity for direct and timely interaction and meaningful dialogue between the Engineer and the successful bidder, prior to execution of a construction contract, are extremely limited and, in many jurisdictions, would violate public competitive bid laws.

In addition, addressing these issues through a value-engineering process (under which the Contractor may propose changes to the design of permanent work) may result in unintended risk assumptions and ambiguity in risk allocation. Similarly, attempts to restrict or control the Contractor's construction means and methods and/or equipment choices through the submittal review process typically introduces ambiguity and uncertainty as to responsibility (Hatem 1998a; Cohen 1995) and results in conflict and disputes between project participants.

In short, there has been no satisfactory resolution in the Design-Bid-Build delivery method for the perceived problems that result from the lack of timely and coordinated integration of the design and construction functions, especially in the context of subsurface projects. Given the intimate and direct correlation between design, construction methodology, equipment selections, and behavior of ground conditions during construction in subsurface projects, these perceived shortcomings have represented major challenges to the process of fairly allocating and managing risk in subsurface projects. How can this "disconnect" between design and construction be addressed and adequately reconciled in the Design-Build process?

The Design-Build method, which, by definition, contemplates the integration of the design and construction functions, could provide an excellent context and opportunity in which to (1) address and resolve many, if not all, of the perceived shortcomings in the Design-Bid-Build method; and (2) achieve a fairer contractual understanding (and, hence, greater predictability and certainty) in risk allocation for design inadequacy, construction deficiencies, and subsurface conditions in subsurface projects. However, in the opinion of this author, it would be a serious mistake to ignore the lessons learned and improved contracting practices in the Design-Bid-Build method by arbitrarily allocating all design, construction, and subsurface condition risk to the Design-Builder.

Roles and Expectations of Principal Design-Build Project Participants and Related Risk Allocation Principles

The Design-Build method provides an opportunity to develop and implement innovative risk allocation and contractual approaches capable of recognizing, adapting, and applying the lessons learned from, and addressing and resolving many of the significant perceived shortcomings and frustrations in, the Design-Bid-Build subsurface project experience. Before discussing the optional contractual approaches available, some general discussion about typical roles and expectations of the principal participants involved in Design-Build subsurface projects would be beneficial.

Owner. The Owner establishes a program (the Owner's program) for the project that defines the conceptual or preliminary design criteria and establishes the quality and performance standards for the design and construction of the completed project; the schedule and budget governing the design and construction process; and any other specific governmental, environmental, or political constraints, including so-called abutter and/or other third-party considerations that must be addressed in the design and construction process. The Owner reasonably should expect that the Design-Builder will satisfactorily achieve the requirements of the Owner's program. The Owner may possess some preliminary subsurface information concerning the site, but that information may or may not be relevant to the eventual development and/or finalization of the Design-Builder's design.

The Owner's Engineering Consultant (typically including a geotechnical engineer and civil engineer) on a major subsurface project typically develops some conceptual or preliminary design to

- Establish or support the feasibility, schedule for, and constructability of the Owner program;
- Provide standards or criteria to be used in the selection of the Design-Builder; and
- Monitor the Design-Builder's design development and construction compliance with the requirements of the Owner program.

The Owner generally should expect to have relatively minimal ability to control the design development process and furnish neither complete design documents nor complete subsurface information to the Design-Builder. The Design-Builder is responsible for the design of the permanent project work and for investigation and development of any subsurface information required in connection with the design and construction of the project consistent with the Owner program, as well as responsibility for the design and implementation of construction means and methods, including equipment selections.

In fact, if the Owner is concerned that exercising a controlling role in the subsurface investigation, design development, and design approval process may result in the assumption of responsibility for the adequacy of the investigation and final design, it may specifically limit involvement in those areas (Hatem 1998a). As such, given the Design-Builder's control over design development and construction, the Owner typically expects that the Design-Builder will have single-point responsibility for the adequacy and performance of project design and construction.

Owner's Engineering Consultant. On Design-Build projects that involve a significant subsurface component, the Owner should retain its own Engineering Consultant, typically including a geotechnical and civil engineer, and sometimes a construction management professional (collectively referred to as the Owner's Engineering Consultant) prior to the Design-Build procurement to serve as the Owner's representative. The scope of services provided by the Owner's Engineering Consultant will depend upon the project-specific contractual scope and terms of engagement. However, in general, the range of services typically includes the following:

- Conducting studies to determine the feasibility—cost, schedule, and constructability—of achieving the Owner program requirements;
- Developing preliminary or conceptual design or performance criteria or standards to be included in the Design-Build request for proposal (RFP);

- Preparing technical portions of the Design-Build RFP,
- Providing the Owner with advice and recommendations in the interview and selection of the Design-Builder,
- Interfacing with the Design-Builder in the design development and finalization processes, and
- Serving as the Owner's representative during construction in monitoring the Design-Builder's performance for compliance with the Owner program and the approved design.

The Owner's Engineering Consultant should also be aware that the greater the degree of Owner control over: (1) the nature and extent of the Design-Builder's subsurface investigation; and (2) the definition, development, and approval of the Design-Builder's design, the greater the probability that both the Owner and the Owner's Engineering Consultant will explicitly or implicitly assume some responsibility for the adequacy of that subsurface investigation and design (Hatem 1998a). The Owner's Engineering Consultant, in short, will need to balance its obligation to represent and protect the interests of the Owner in terms of the Design-Builder's satisfaction of the program requirements while, at the same time, not exercising an overly intrusive or dominant role over the various activities of the Design-Builder. In attempting to limit its potential professional liability exposure in representing the interests of the Owner, the Owner's Engineering Consultant will often point to its limited authority, especially relating to design development and defining the scope of and adequacy of the subsurface investigation program.

Design-Builder. The Design-Builder reasonably should expect to have a fair degree of flexibility and discretion in the development of a design that will satisfy the Owner program requirements and, in so doing, achieve a reasonable profit. In responding to the RFP, the Design-Builder will want (and should have available to it) any subsurface information and preliminary, conceptual design information that is available. The Design-Builder will realize that since it is responsible for developing and finalizing the design, it should be responsible for defining and undertaking a subsurface investigation program required to support that design and to address constructability considerations or other aspects required in connection with developing and constructing the design consistent with the Owner program requirements. The Design-Builder should be afforded a reasonable amount of time and a meaningful opportunity to investigate site conditions and review preliminary design and/or site information furnished by the Owner. Efforts by the Owner and/or the Owner's Engineering Consultant to significantly limit or constrain the Design-Builder's flexibility and discretion, such as through the imposition of detailed and prescriptive design criteria or performance standards or stringent submittal review and approval requirements, will be perceived by the Design-Builder as an intrusion into the Design-Builder's role, which should result in some risk retention to the Owner for the adequacy of final design and subsurface investigation.

The engineers (geotechnical, civil, or both) who are part of the Design-Build team (collectively the Design-Build Engineer) similarly will want to have a reasonable degree of discretion and flexibility in the development and finalization of the design and the definition of the subsurface investigation program. The Design-Build Engineer reasonably will want a scope of engagement that is adequate to allow for design development and finalization, as well as to be able to identify and evaluate in the construction phase whether encountered conditions require a reconsideration of prior design assumptions, approvals, and decisions (Hatem 1998a). In circumstances in which the Owner and/or the Owner's Engineering Consultant has established design criteria that must be followed in the Design-Builder's design development effort, the Design-Build Engineer should be afforded adequate time and a meaningful opportunity to satisfy itself as to the adequacy and appropriateness of those criteria since, in the final analysis, the Design-Build Engineer will be responsible for the adequacy of the final design (Hatem 1998a).

Finally, the Design-Builder will endeavor to include in the Design-Build agreement a differing site condition (DSC) clause and some alternative dispute resolution mechanism, such as a dispute review board (DRB), for the determination of DSC disputes that fall within the scope of that clause. The combined inclusion of a DSC clause and a DRB process will function best when the Design-Build agreement objectively defines the baseline physical site conditions anticipated and forms the contractual basis of the agreement so as to facilitate a determination whether the conditions actually encountered are substantially or materially different from those contractually defined or assumed.

SUBSURFACE RISK

This section will address the various contractual approaches that may be chosen by an Owner to allocate risk between the Owner and the Design-Builder for subsurface conditions on a Design-Build project, and the secondary implications of those choices on the professional liability of Engineers who are part of the Design-Build team, as well as those who serve as the Owner's independent consultants. Unquestionably, Owners are the principal decision-makers regarding risk allocation for subsurface conditions. This is an important decision in a subsurface project, and the Owner has many options. As stated by knowledgeable commentators:

> For tunneling and underground construction, perhaps the most important element of the project for which risk and responsibility must be clearly allocated is subsurface conditions . . . [I]n Design-Build, the Owner may perceive the opportunity to allocate additional risk for subsurface conditions to the Design-Build team. This approach could result in the level of advance work ranging from the execution and presentation of a thorough subsurface investigation and limited interpretation, to incomplete or non-existent exploration and no interpretation. (Essex and Zelenko 1999)

Engineers serve a critically important consulting role in advising Owners (and sometimes the Owner's legal counsel) regarding the appropriate factors to be considered in deciding among these various risk allocation approaches (Hatem 1998a). Owner decisions represent significant project planning decisions, especially in those projects involving a substantial subsurface construction component. Those decisions will impact a number of project considerations beyond strictly risk allocation, such as

- Relationships among project participants,
- Likelihood of disappointed performance and commercial expectations and disputes,
- Ability of the Contractor to deliver the project on time and budget,
- Ability of the Contractor to complete the project and make a reasonable profit,
- Ability of the Owner to obtain an adequate pool of qualified bidders, and
- Desirability of procuring a bid that accurately reflects the real cost of the work without excessive contingencies for ill-defined risks.

How can these considerations be reconciled, balanced, and managed so as to result in a fair and reasonable process of risk allocation for subsurface conditions among the principal project participants on a Design-Build project? This process should take into account the differences between the roles and expectations of the participants in Design-Build and those in Design-Bid-Build. In addition, the process should strive to adapt and enhance the improved contracting practices, especially given the ability in Design-Build to overcome many of the perceived short-comings and frustrations of Design-Bid-Build in the context of subsurface projects. There are two primary contractual approaches on a Design-Build project that may be considered.

Transfer All Risk to the Design-Builder. An Owner may elect to transfer to the Design-Builder all risk for subsurface conditions. Under this contractual approach, the Owner may provide the Design-Builder with little or no subsurface information or may provide subsurface information but disclaim the accuracy and completeness of that information and, additionally, may choose not to include a DSC clause in the Design-Build agreement. An Owner choosing this all risk to the Design-Builder contractual approach may employ the following line of reasoning:

- The Design-Builder has single-point responsibility for the design and construction of the project consistent with the Owner program requirements.

- In contrast to the Design-Bid-Build method, the Owner does not furnish a detailed or completed design to the Design-Builder, and the latter is responsible for the development and finalization of design and, hence, for the adequacy of the design consistent with the Owner's program requirements.

- The nature and extent of subsurface investigation is inextricably tied to the design approach, decision-making, and final design. The party responsible for the final design—the Design-Builder—should, therefore, be responsible for the adequacy of the scope, definition, and undertaking of the subsurface investigation program.

- The Design-Builder's construction means and methods and equipment may be influenced by information concerning the anticipated subsurface conditions and, accordingly, since the Design-Builder is responsible for the former matters, it should also be responsible for the subsurface investigation.

- Because the Design-Builder will have single-point responsibility for design and construction, it is entirely consistent that the Design-Builder be responsible for any cost and time impacts associated with subsurface conditions. Hence, if an Owner retains any level of responsibility for subsurface information, preliminary design criteria, or performance standards that the Owner may have furnished to the Design-Builder, or the inclusion of a DSC clause in the Design-Build agreement, then such retention would subvert and undermine the otherwise clear intent and expectation to allocate all risk to the Design-Builder due to the inadequacy of the subsurface investigation program or the design.

The all-risk approach, while superficially logical in relation to the Design-Build principle of single-point responsibility, is not recommended as it has the following negative project impacts:

- The all-risk approach will undermine the bidding process by encouraging risk gambling, by reducing the pool of qualified and responsible bidders, and by planting the seeds for seriously disappointed performance and commercial expectations of the parties. Also, in some instances, the protections afforded by the absence of a DSC clause may be illusory if courts allow relief to the Contractor under one or more equitable or legal theories. In addition, depending upon the level of design detail furnished to the Design-Builder, and the relationship of that design to the subsurface investigation required to support the design, a court may conclude that, despite the absence of a DSC clause, the Owner is responsible for the time/cost implications of subsurface conditions, given the detail and prescriptive nature of the design furnished by the Owner (Hatem 1998a).

- The Owner may have greater exposure to third-party claims and considerations (e.g., adjacent property owners and regulatory authorities), as well as to adverse public relations for the project.

- The quality of the completed and permanent subsurface project work may be compromised.

- In substance, the Design-Builder assumes all risk of time/cost impact arising out of differing site conditions, consistent with the common law rule. Depending upon the nature and extent of the opportunity to conduct subsurface investigation prior to contract execution, the Design-Builder may assume a significant unquantifiable and contingent risk.

- From the standpoint of both the Owner's Engineering Consultant and the Design-Build Engineer, the risk of professional liability claims arising out of subsurface conditions significantly increases given the absence of any contractual remedy to the Design-Builder due to the encountering of subsurface conditions materially or substantially different from the Design-Builder's contractual assumptions and reasonable expectations (Hatem 1998a).

Risk-Sharing Approach.   Suffice it to say, there is an emerging debate as to how subsurface condition risk should be allocated on Design-Build projects. However, the commentary to date, and with good reason in the context of risk allocation for subsurface conditions on Design-Build projects, supports and endorses a risk-sharing approach accomplished through the use and adaptation of the improved contracting practices.

Assuming the inclusion of a DSC clause as a risk-sharing mechanism on a Design-Build project, how may risk for subsurface conditions be more precisely, objectively, and fairly allocated between Owner and Design-Builder? Conscientious consideration of this question requires a balancing of the differing roles and responsibilities of Owner and Contractor in the Design-Bid-Build method with those of Owner and Design-Builder in the Design-Build method with respect to: (1) responsibility for design of the completed work; and (2) the opportunity to define and undertake the subsurface investigation program. As discussed, these distinctions have important risk allocation and professional liability implications for participants in the Design-Build process. In exploring this question, it has been stated:

> Shifting the risk of unanticipated site conditions [DSC] to the Owner also makes sense in Design-Build project delivery, but the issues involved are somewhat different from those in Design-Bid-Build contracting. Depending upon when the Design-Builder is retained, the Design-Builder may have significant responsibility for predesign site evaluation and may be assigned the responsibility for developing the geotechnical investigation program. This responsibility, coupled with the fact that the Design-Builder develops the design and drafts the construction documents, mandates that parties consider what site condition risks the Owner retains and what site condition risks are transferred.
>
> One way to address site condition risks in Design-Build contracting is for the Owner to establish during the development of its concept documents and design criteria a listing of geotechnical assumptions based upon either a preliminary site exploration program or information from previous building programs. This information is then considered a baseline for the Design-Builder to rely on that, if incorrect, triggers the application of the "Differing Site Conditions" clause. Alternatively, the Owner and the Design-Builder can agree upon an investigation program that will be used as the baseline for differing site conditions claims. Regardless of which method is chosen, the standard "Differing Site

Conditions" clause must be modified to reflect the different circumstances. (Loulakis and Shean 1996)

The important distinctions between the roles in Design-Bid-Build and in Design-Build certainly support the conclusion that, in Design-Build, the Design-Builder should assume significantly greater responsibility for the risks associated with the adequacy of both subsurface investigation and the design of the completed project. In the opinion of this author, despite the greater responsibility of the Design-Builder, the all-risk approach to allocation of risk for subsurface conditions is neither fair nor appropriate, and some modified risk allocation approach based on risk sharing is preferred.

How and where to draw the line in the allocation or transfer of risk poses a challenge but also an opportunity. The challenge is how to allocate risk between Owner and Design-Builder in a manner that adequately and fairly accounts for the increased risk appropriately allocated to the Design-Builder, given the latter's greater involvement, direction, control, and decision-making with respect to subsurface investigation, design development, and finalization processes. At the same time, an important goal of both Owner and Design-Builder is to allocate risk in a manner that accurately, objectively, and clearly evidences and documents the expectations of those parties at the point of contract formation.

As previously noted, the Design-Build method affords the opportunity to achieve those objectives without many of the constraints, perceived shortcomings, and frustrations inherent in the Design-Bid-Build method. As a part of the contract negotiation and formation process on a Design-Build project, the Owner and Design-Builder may mutually agree upon a program of subsurface investigation adequate and appropriate for the Design-Builder's proposed design approach to the project. The Design-Builder may be responsible for undertaking that subsurface investigation program and, ultimately and contractually, would be responsible for the adequacy of that program to support its proposed design approach. The Owner, typically through its Engineering Consultant, would have the ability and opportunity to monitor the Design-Builder's compliance with the mutually agreed-upon subsurface investigation program so as to represent the Owner's interest as to the adequacy and quality of the investigation and otherwise with respect to the achievement of the Owner program.

Upon the Design-Builder's completion of the subsurface investigation, and prior to the point of contract formation, the Owner and the Design-Builder are in a prime position to establish a record of negotiation and agreed-upon project-specific subsurface baseline conditions or assumptions upon which their contractual expectations and the Design-Build agreement are predicated. At that point, a DSC clause may be developed that specifically defines, in relevant project-specific parameters, the Design-Builder's risk assumption as including all site and subsurface condition risk within the contractually defined or assumed baseline parameters, and allowing for risk transfer to the Owner should subsurface conditions be materially or substantially different from the baseline conditions encountered during performance of the work.

The baseline approach to contracting on subsurface projects is discussed in detail in *Geotechnical Baseline Reports for Underground Construction—Guidelines and Practices, Technical Committee on Geotechnical Reports of the Underground Technology Research Council*, by R.J. Essex (Essex 1997). Although most of the published literature to date discusses the application of the baseline approach in the context of Design-Bid-Build subsurface projects, that approach has significant potential for constructive and innovative application in the Design-Build delivery method. Generally speaking, the baseline approach is utilized on subsurface projects to establish a contractual statement of the geotechnical conditions anticipated during underground and subsurface construction. These baselines establish a contractual basis for the allocation of geotechnical risk and are not necessarily intended to define the single correct interpretation of geotechnical conditions.

In substance, the purpose of the baseline approach is to assist in the allocation of risk by facilitating the definition—qualitatively and/or quantitatively—of those site conditions that should be anticipated and included within the contractual obligations and contract price of the Contractor. The baseline approach thereby reduces the likelihood of misunderstandings, disappointed expectations, disputes, and claims (and associated project delays and disruptions) during the construction process. In addition, the baseline approach may also be utilized (with appropriate care) in Design-Build in the context of risk allocation for the cost/time impacts due to variations in contractually assumed or expected construction means and methods and/or equipment selections resulting from unanticipated subsurface conditions.

The Tren Urbano Project in Puerto Rico provides an illustration of the application of the baseline approach as a contractual method on a major Design-Build subsurface project. The contract documents for the project included several key components critical to the definition and implementation of risk allocation for subsurface conditions:

- Baseline assumptions regarding critical subsurface conditions,
- DSC clause triggered by a variation from the baseline assumptions and the encountered conditions, and
- DSC clause to provide a mechanism for an equitable adjustment to the Design-Builder due to any such variations.

In addition, the Design-Build agreement on the Tren Urbano Project qualifies the Design-Builder's responsibility for subsidence claims and damages based upon the Design-Builder's satisfaction of certain baseline criteria regarding assumed and contractually defined construction means and methods. Simply put, the Tren Urbano project represents a good illustration of an Owner's attempt to achieve a risk-sharing approach for subsurface conditions on a Design-Build subsurface project.

Assuming that an Owner is willing to implement the risk-sharing approach for subsurface conditions on a Design-Build project, the following are the guidelines that should be considered:

1. The goal is for the Owner and Design-Builder to arrive at an objectively and clearly defined threshold or, more appropriately, a series of thresholds, that establish—for project-specific contractual risk allocation purposes—the point of subsurface condition risk transfer from one party to the other. These thresholds may be established as a result of a collaborative effort by the Owner (typically, with the assistance of its Engineering Consultant) and the Design-Builder (including members of the Design-Builder's engineering design team). This collaborative effort would involve discussions and interaction between the Owner's Engineering Consultant and the Design-Builder in which information and ideas would be shared both ways regarding anticipated site conditions, design concepts and development, constructability considerations, construction means and methods, equipment selections, and third-party considerations, as well as other factors relevant to the assessment and management of risk and the successful design and construction of a subsurface project consistent with quality and other objectives and requirements of the Owner program.

2. The Owner and the Owner's Engineering Consultant may be concerned that, by becoming involved with the Design-Builder in such a collaborative and interactive effort in

establishing the baselines, they may assume too much responsibility for successful completion of the project. This is a legitimate concern and, for that reason, it is advisable that the Design-Build agreement state that neither the Owner's issuance of the baselines, nor any of the recommendations, input, or performance of services of the Owner's Engineering Consultant in the negotiation and/or establishment of the baseline shall relieve the Design-Builder of its exclusive responsibility for the adequacy, accuracy, and completeness of the subsurface investigation program, the design of permanent work, and the design and implementation of construction means and methods and safety precautions and programs.

3. In accomplishing this goal, it is important that the thresholds established be incorporated in the Design-Build agreement either as baselines or baseline assumptions. In addition, consideration should be given as to how the baselines would operate in relation to the DSC clause contained in the Design-Build agreement. In other words, will *any* variation between the baseline and the physical site conditions actually encountered trigger an equitable adjustment under the DSC clause, or must the variation be *material* or *substantial* in nature (Hatem 1997)? There are other issues that should be addressed in the development and contractual articulation of baselines, as well as professional liability implications, that are beyond the scope of this chapter, but addressed elsewhere (Hatem 1997).

4. Consistent with the general risk allocation principles underlying the Design-Build method, it would reasonably be expected that the Design-Builder would be responsible for

   • Development, accuracy, and completeness of its own subsurface investigation program required to support the design and construction of the project;

   • Adequacy and completeness of the design of the permanent work; and

   • Design and implementation of construction means and methods, equipment selections, and safety precautions and programs.

RISK ALLOCATION FOR DEFECTIVE PRELIMINARY AND FINAL DESIGN

The increasing popularity and utilization of the Design-Build method pose new and important challenges to the process of clearly and fairly allocating risk among project participants. Many Owners select the Design-Build method in order to transfer or allocate to the Design-Builder all or the predominant risk for both design and construction defects—that is, single-point responsibility. Many Design-Builders are willing to contractually assume that responsibility, provided they are given a commensurate degree of discretion and control over the process of developing and finalizing the design and are otherwise in control of the variables affecting that degree of risk assumption. At the same time, however, many Owners seek to limit that discretion and control by furnishing to the Design-Builder mandatory preliminary design criteria and/or performance specifications that the Design-Builder must follow in the design development and finalization process. Moreover, many Owners exert yet further influence and control over the Design-Builder's design development and finalization process by requiring their approval (subject to more or less defined standards) of progress and final design submissions during that process.

Who should bear the risk and responsibility for defects in Owner-furnished preliminary design and/or performance specifications? Are some Owners seeking the best of both worlds—controlling the Design-Builder's design development and finalization process without being responsible for defects in the preliminary or final design? Is the distinction between responsibility for preliminary design and final design one that realistically can or should be drawn, and, even if so, is it a sensible and practicable distinction in the Design-Build context? Is there a risk allocation approach for defects in the preliminary design (Owner-furnished) and final design (Design-Builder–furnished) that is fair, practical, and makes sense from the professional registration and legal perspectives?

All of these issues are important in the specific context of Design-Build subsurface projects, especially given the interrelationship between the final design and (1) the nature and extent of subsurface investigations; (2) the construction means and methods, and equipment selections made in constructing the project; and (3) the behavior or other response of subsurface conditions during construction. Risks, roles, and responsibilities for these various interdependencies may be shared among Design-Build project participants or severally and exclusively assigned to individual participants. The ability to control the adequacy of the final design may well depend upon how those roles are allocated. As in the case of subsurface condition risk allocation, the Owner is the principal decision-maker regarding these options, and its Engineering Consultant serves a critically important advisory role.

In the Design-Build method, the Owner typically provides the Design-Builder with conceptual criteria, preliminary design, and/or performance specifications, which the Design-Builder relies upon in responding to the Owner's RFP and is required to utilize in developing and finalizing the project design and in constructing the project so as to achieve the Owner's programmatic goals. Whether explicitly or implicitly acknowledged in the Design-Build agreement, some fundamental and important issues are presented in this context regarding allocation of risk and responsibility for the Owner-furnished or preliminary design and performance specifications.

Unless these issues are explicitly identified and adequately addressed in the Design-Build agreement, there is a substantial probability that the allocation of responsibility for the final project design, and the ability of that design to achieve performance requirements, will be unclear and unduly difficult to ascertain. The resulting fragmentation and diffusion of responsibility for defective design among participants undermines one of the principal goals of Design-Build—to accomplish unified or single point of accountability for the design, construction, and performance of the completed project.

In the process of evaluating information contained in the RFP (or other procurement solicitation or tender), and in preparing its response or proposal, the Design-Builder may or may not have adequate opportunity to independently verify the adequacy, accuracy, or constructability of the Owner-furnished preliminary design, or the ability to achieve the Owner's performance specifications. However, the RFP and the eventual terms of the Design-Build agreement typically require that the Design-Builder *ultimately* be exclusively responsible for both the preliminary and the final design, and may even require that the Design-Builder provide warranties regarding the adequacy of that design. Under such ultimate obligations, the Design-Builder typically assumes risk and responsibility for any Owner-furnished preliminary design and/or the achievability of Owner-furnished performance specifications.

A compelling position may be advanced that for all purposes—including civil (contractual) liability as well as registration law compliance—the Design-Builder (or, more precisely its Engineer who prepares and stamps the final design) should be exclusively and completely responsible for the final design as well as for the preliminary design, which (even though Owner-furnished) forms the underlying and developmental basis of the final design. This approach to risk allocation for defective or inadequate design is consistent with the hallmark Design-Build single-point responsibility principle and also promotes the general public policy objective of avoiding fragmentation and diffusion of design responsibility. It certainly should serve to mini-mize disputes and claims pertaining to design adequacy issues because the lines of responsibility are absolutely clear, unqualified, and exclusive (fairness aside, at least for the moment).

An alternate approach—presumably grounded in a sense of fairness—is to share or differ-entially allocate design defect risk between the Owner and Design-Builder by allocating (1) to the Owner responsibility for the Owner-furnished preliminary design and/or performance speci-fications; and (2) to the Design-Builder final design adequacy and performance responsibility, subject to the qualification that the Design-Builder will not be responsible for inadequacies or defects in the Owner-furnished preliminary design or performance specifications.

The shared allocation of risk approach superficially appears reasonable based upon which (the Owner or Design-Builder) is responsible for the levels of design preparation. However, defin-ing responsibility along such lines, from a professional liability, registration law, and pragmatic approach, inevitably leads to disputes and claims between the Owner and Design-Builder and, perhaps most importantly, serves to undermine the fundamental and essential single-point prin-ciple that is the cornerstone of the Design-Build delivery method.

Both of these alternate approaches for defective preliminary and final design—while signifi-cantly different—contemplate that risk allocation for both preliminary and final design be made *at the point of contract formation*—that is, at the execution of the Design-Build agreement.

Despite the legal and other logic, an elemental sense of fairness militates in favor of afford-ing the Design-Builder a reasonable opportunity to verify the adequacy of the Owner-furnished preliminary design and/or the feasibility of achieving the Owner's performance specification requirements *prior to* the Design-Builder undertaking contractual (or warranty) responsibility for the final design. In many instances, however, the Design-Builder is not afforded and does not reasonably have such an opportunity *prior to* execution of the Design-Build agreement. In fact, often the opportunity realistically and reasonably does not exist until *after* the Design-Builder has executed the Design-Build agreement and has had the opportunity to undertake and accomplish a reasonable degree of evaluation, validation, feasibility review, and development of the Owner-furnished preliminary design and performance requirements.

Risk Allocation for Defective Preliminary Design and Performance Specifications

The objective of allocating to one entity—the Design-Builder—single-point accountability and responsibility for the design and construction is attractive to many project Owners and one of the most important factors militating in favor of its selection. The extent or degree of risk allocation in Design-Build varies and primarily is dependent upon the Owner's preferences and the product (more or less) of a negotiation process with the Design-Builder. However, risk allocation need not be an all-or-nothing proposition, and the Owner has a number of options available to quali-tatively and quantitatively allocate and share defective design risk in the Design-Build approach, as well as to time or sequence the transfer of any such risk allocation or sharing.

At the end of the day, most Owners utilizing the Design-Build method want the Design-Builder to be responsible for the final design, including the underlying preliminary design. As an *ultimate* objective, this risk allocation approach seems reasonable, especially given that the Design-Builder is responsible for development and finalization of the design, retains its own Engineer, and has the ability to control the design development and finalization process. However, prior to and during that process, and beyond initially furnishing preliminary design and performance specifications, there may be varying degrees of involvement by the Owner (and/or its Engineer-ing Consultant) that typically will further influence the design concept, limit discretion of the

Design-Builder in the development of the design, and impact the achievability of the Owner's project performance requirements.

More specifically, the terms of the Design-Build agreement usually define the Owner-furnished (or mandated) preliminary design and performance specification requirements that the Design-Builder must adhere to and achieve, as well as the degree of discretion that the Design-Builder will have in developing the design and the extent of the authority of the Owner (and/or its Engineering Consultants) in reviewing and/or approving design development submittals of the Design-Builder. Thus, even though the Design-Builder ultimately will be responsible for the adequacy and performance of the final design, in the process leading to that end result, the Owner (and/or its Engineering Consultant) often has (have) significant influence in developing the conceptual underpinning of that design and some degree of control over the design development and finalization process.

The Owner's ultimate objective of risk transference to the Design-Builder for final project design and performance must be reconciled with the Owner's legitimate interests in providing to the Design-Builder preliminary design and performance specifications. These interests may generally be identified as follows:

- The Owner may have specific project design criteria or standards, or performance objectives that it must follow or adhere to in order to meet its internal programmatic or other requirements.

- The Owner may have developed preliminary design as a basis to estimate and establish project cost and funding commitments.

- The Owner may be required to satisfy specific design criteria or performance standards imposed by third parties, regulatory authorities, or funding sources.

- The Owner may have invested in the cost of preliminary design development prior to making a decision whether to utilize the Design-Build method.

- The Owner's preliminary design and/or performance specifications may be the basis or condition of project permits or approvals previously granted by, or negotiated with, regulatory agencies, funding sources, ultimate project end-users, or other third parties having an interest in or affected by the project.

- The Owner may wish to utilize the preliminary design and performance specifications as

  - Baseline or framework by which to evaluate responses from various Design-Builders responding to the RFP or other procurement process and as a basis to select the Design-Builder;

  - Basis to define risk allocation, warranty, performance standards, and other provisions of the Design-Build agreement;

  - Objective basis to evaluate the adequacy of the Design-Builder's design submissions and performance required under the Design-Build agreement; and

  - Evaluation of claims and/or as a basis for resolving disputes with the Design-Builder for adjustment of contractually allocated risk, cost, and time obligations.

Discussions of some general definitions and their risk implications are appropriate at this point. By *performance specifications*, an Owner sets forth objectives to be achieved, standards to be attained, or parameters to be followed, with the intention that the Design-Builder be given significant discretion, flexibility, and latitude in accomplishing the broad objectives (explicit or implicit) stated in the performance specifications.

A *pure performance specification* gives the Design-Builder virtually complete control and discretion over the ultimate design, appearance, functionality, and configuration of the finished product, along with a corresponding (heightened) assumption of risk and responsibility for that design. An Owner (or Engineering Consultant) preparing a performance specification should have in mind

- Purpose of the specification;
- Need to provide the Design-Builder with a reasonable degree of design discretion in exchange for the latter's assumption of defective design risk; and
- Goal of achieving quality in the process of developing a specification that adequately addresses, balances, and defines the following:
  - Owner's budget and program
  - Schedule
  - Project complexity and interface requirements
  - Performance requirements or expectations
  - Levels of acceptable finish and quality
  - Project expansion requirements
  - Aesthetic requirements
  - End-user requirements.

In utilizing a *conceptual* or *preliminary design* approach, the Owner—often through an Engineering Consultant—typically develops the design to the 20% to 35% level, with the intention and expectation that the Design-Builder will be responsible for developing and finalizing the design consistent with that preliminary design. In general terms, use of the preliminary design approach results in a more restricted discretion of the Design-Builder than the pure performance specification approach.

These definitional distinctions between performance specifications and preliminary design, along with the corresponding and relative risk allocation implications for the Owner and Design-Builder, may be stated with superficial clarity. However, in application, these distinctions may be blurred during project execution, resulting in ambiguity in risk allocation. In general terms, use of preliminary design and performance specifications are intended to confer upon the Design-Builder substantial discretion and latitude for design development and, hence, a corresponding (heightened) degree of risk transference to the Design-Builder for the final design.

Causes of Defective Preliminary Design. Some examples of defects in preliminary design and/or performance specifications furnished by the Owner to the Design-Builder include

- Errors or omissions;
- Conflicts and inconsistencies between preliminary design, performance specifications, and other mandatory criteria or standards furnished and required by the Owner;
- Impossibility or impracticability of achieving required performance standards or requirements;
- Conflicts or inconsistencies between preliminary design and standards required by applicable code or governmental or regulatory authority; and

- Incompatibility of preliminary design with project geography, environmental, or other project or site-specific constraints.

As a general matter, an Owner furnishing preliminary design and/or performance specifications to the Design-Builder typically will state in the RFP whether

- The Design-Builder is entitled to rely upon the accuracy, adequacy, or constructability of the preliminary design and/or performance specifications, or whether the Design-Builder must independently verify and be responsible for same; and/or
- The accuracy, adequacy, and/or achievability of the preliminary design/performance specification is disclaimed.

The failure to explicitly address risk allocation for deficiencies in Owner-furnished preliminary design or performance specifications in all probability will increase the possibility of disappointed expectations, and associated and ensuing disputes and claims between the Owner and Design-Builder. Some Owners may consciously focus on this risk allocation issue but elect not to specifically address it on the assumption that, because the Design-Builder will be required to warrant and otherwise be responsible for the adequacy of the final design, the latter's ultimate responsibility necessarily and by implication includes responsibility for the underlying preliminary design and/or performance specifications upon which the final design is predicated. Despite the logic of that implication, it is strongly advisable to specifically and explicitly address in the Design-Build agreement the issue of risk allocation for defects in Owner-furnished preliminary design and/or performance specifications.

This issue is more complex than may superficially appear to be the case. The development and convergence of the Owner-furnished preliminary design into the Design-Builder's final design; the Design-Builder's expected ultimate responsibility for the final design; and the contractual, actual, and respective roles of the Owner, its Engineering Consultant, and the Design-Builder and its Engineer all combine and serve to challenge and complicate the process of clearly defining and allocating risk and responsibility between Owner and Design-Builder (and their respective Engineering Consultants) for the preliminary and final designs. However, assuming that one could do this, one would need to address the professional standards required or expected with the preparation of preliminary design and final design. Put another way, what constitutes defective design in the preliminary and final design context?

In the absence of an explicit statement in the RFP or Design-Build agreement, it reasonably should be expected that the professional standard of care would apply. This standard has been defined as follows:

> Architect, doctors, engineers, attorneys, and others deal in somewhat inexact sciences and are continually called upon to exercise their skilled judgment in order to anticipate and provide for random factors which are incapable of precise measurement. The indeterminable nature of these factors makes it impossible for professional service people to gauge them with complete accuracy in every instance. Because of the inescapable possibility of error which inheres in the services, the law has traditionally required, not perfect results, but rather, the exercise of that skill and judgment which can be reasonably expected from similarly situated professionals. (*Klein v. Catalano* 1982)

This standard does not require perfection and is flexible in its application, as well as highly fact- and circumstance-dependent. Accepting that preliminary design, by definition, is incomplete and based upon limited information and developmental effort, significant considerations must be taken in defining the acceptable standard of care governing it and in evaluating responsibility for its adequacy. In addition, preliminary design is prepared in the expectation that the design will be further questioned, validated, developed, and finalized by a qualified design professional who, ultimately, will stamp the final design and that, during this design development process, the design will undergo constructability and related reviews by the Design-Builder, which will be responsible for the design and construction of the project.

The evolving case law regarding design risk and responsibility in the Design-Build method supports the view that the Owner may be liable to the Design-Builder for (1) erroneous or incomplete information; and/or (2) defective preliminary design, furnished to and reasonably relied upon by the Design-Builder in developing and finalizing the project design.

For example, in *Pitt-Des Moines, Inc.*, the Board of Contract Appeals decided in favor of the Design-Builder in finding that the Owner failed to provide the Design-Builder with all information available and provided inaccurate information about existing conditions at the project site. While the Owner pointed to a disclaimer clause in an attempt to absolve itself of liability, the board did not find the Owner's argument convincing.

In *Pitt-Des Moines*, the Design-Build agreement required the Design-Builder to undertake certain structural modifications to an existing building. In preparing its technical proposal, the Design-Builder reviewed the RFP drawings, inspected the site, and attended the preproposal conference. Since the RFP drawings were illegible and vague, the Design-Builder requested that the Owner provide additional drawings, but the Owner informed the Design-Builder that no other drawings were available. At the preproposal conference, the Owner also assured the participants that there were no known problems with the adjacent building.

Based upon its review of the RFP drawings, its inspection of the site, and its reliance on the Owner's assurances, the Design-Builder proposed a type of piling system required under the building. After its proposal was submitted, the Design-Builder discovered that the RFP drawings were inaccurate and that, contrary to the Owner's earlier assurance, additional drawings existed. The result was that the Design-Builder concluded that a complete and costly redesign of the pile system was required.

The Design-Builder asserted a DSC claim, and the Board of Contract Appeals agreed that the RFP drawings were inaccurate. The Owner argued that because the Design-Builder signed a Design-Build agreement, the Design-Builder assumed greater responsibility for site conditions, including the obligation to field check the RFP drawings, and that the Design-Builder acted unreasonably in failing to do so. The Owner supported its argument by highlighting the contract's disclaimer about the RFP drawings that read, in part, that the drawings "may not be accurate and shall be field-checked."

In its ruling for the Design-Builder, the board stated that the Design-Builder was reasonable in relying upon the Owner's assurances. "Potential Constructors are required to take all steps reasonably necessary to ascertain the nature and location of the work, and to satisfy themselves as to the general and local conditions affecting the work. However [the Contractor] is not required to conduct a costly or time consuming technical investigation to determine the accuracy of . . . drawings." The board found that the Design-Builder had met this standard by its repeated questioning of the Owner and had made reasonable efforts to investigate the site. Furthermore, the Owner had an affirmative duty to provide any relevant information about the pilings and the structural work to be performed.

The board also determined that the interpretation and purpose behind the DSC clause is the same whether it is written into fixed-price construction contracts or Design-Build contracts. The philosophy in both instances is that bidders must feel confident that they can rely on information provided by the Owner in making their proposals. To permit otherwise would cause bidders to establish large contingencies, resulting in potentially inflated bids.

Similarly, defective preliminary design furnished by the Owner to the Design-Builder may create liability exposure for the Owner. In *M.A. Mortenson Co.*, the Board of Contract Appeals found for the Design-Builder in stating that the Owner had a duty to provide accurate conceptual drawings that were to be the basis for the Design-Builder's proposal. In *Mortenson*, the Design-Builder had contracted to design and construct a medical clinic at an Air Force base. In preparing its bid for the project, including its proposal for the cost of quantities of structural concrete and reinforcing steel needed, the Design-Builder relied on the 35% complete conceptual drawings that the Owner supplied to all bidders because the RFP itself stated that the drawings "may be used to form the basis for the pricing proposal." The scope of work required that the Design-Builder "verify and validate the accuracy of the preliminary design information and submit complete design documents."

The final design required more concrete and steel than was indicated in the Owner's conceptual design documents or the Design-Builder's bid proposal pricing. Given the significant cost of the increased quantity required, the Design-Builder presented a claim based on the contract's changes clause. The Owner argued that the Design-Builder should not have relied on the conceptual drawings without retaining a structural engineer to review them and should have included a contingency in its pricing to cover any potential increase in concrete and steel quantities.

Like the *Pitt Des-Moines* case, the *M.A. Mortenson* board also applied a Design-Bid-Build standard to this Design-Build scenario. The board paid no heed to the philosophy of single-point responsibility in the Design-Build method and the corresponding release of liability of the Owner. To the contrary, the board viewed the claim from the perspective of a reasonably prudent Design-Builder and deemed it unreasonable for a Design-Builder to have to perform a thorough investigation at so early in the process: "The contract required . . . [the Design-Builder] to verify and validate the design as part of the *design* work, not the *proposal* effort" (emphases added). In addition, the board pointed out that if it were to shift the burden to the Design-Builder, it would completely negate the purpose of the changes clause.

These two cases illustrate that the Owner may not be able to shift all responsibility for failing to provide accurate or complete information and/or preliminary design to the Design-Builder. *Pitt Des-Moines* emphasizes the importance of the Owner disclosing all material information that may affect the Design-Builder's work and providing the most accurate information available to the Design-Builder. *M.A. Mortenson Co.* emphasizes the importance of the Owner making every reasonable effort to provide accurate preliminary design to the Design-Builder. In addition, the Design-Builder's Engineer may be subject to liability for the cost of work indicated in the RFP but not included in design or related information furnished by that consultant to the Design-Builder in the preparation of the latter's pricing or other response to the RFP.

Risk Allocation for Defective Final Design.   In the Design-Build method, the Design-Builder has responsibility to prepare and stamp (by its Engineer) the final design. Given that responsibility, the expectations of the Owner and Design-Builder and the corresponding terms of the Design-Build agreement typically contemplate that the Design-Builder will be allocated the risk of any inadequacies or other deficiencies in the final design and will have ultimate responsibility for the performance of that final design.

However, having stated the general role and responsibility of the Design-Builder, it is also important to emphasize that the Owner typically furnishes to the Design-Builder preliminary design (with varying degrees of completeness) and/or performance specifications, which the latter is required to utilize and follow in the design development and finalization process. During the latter process, the Owner and its Engineering Consultant may require design development submissions from the Design-Builder and have a role in the review and approval of the latter design submissions. Thus, while the expectation that final design risk and responsibility ultimately will be allocated to the Design-Builder is entirely consistent with the single-point principle and the corollary principle that the party in control of the design process should bear the risk of inadequacies that result from that process, the nature and extent of the role of the Owner (and/or its Engineering Consultant) in the preliminary design and in the review and approval of the Design-Builder's design development submissions must be taken into account and balanced in the allocation of risk and liability for defective design.

The *degree* of development of the preliminary design may vary from an average of 20% to 35% on the lower end to a higher range of 60% to 70%. As a general proposition, the lower end confers greater discretion, flexibility, and control upon the Design-Builder to develop and finalize the design, while the higher range confers significantly less. In other similar contexts, courts have recognized that when final design preparation and responsibility is delegated to a Contractor based upon a preliminary design that is substantially complete when furnished (leaving only detailing for the Contractor) or when the discretion and control of the Contractor (and its design professional) is significantly curtailed or limited (through, for example, a dominant and unconstrained Owner review/approval process), the Owner (and its Engineering Consultant) may retain some shared or all responsibility for the adequacy of the final design, even though the final design responsibility was formally or contractually delegated to the Contractor.

Similarly, even when the degree of preliminary design preparation is on the lower end of the range and the Design-Builder (and its Engineer) has adequate control in the process, an Owner (despite provisions in the Design-Build agreement allocating final design risk and responsibility to the Design-Builder) may reassume defective design risk by exercising a dominant or intrusive role, directly or through its Engineering Consultant, in imposing or directing changes or other design initiatives or limitations in the process of reviewing and/or approving the Design-Builder's design submissions, or when an Owner otherwise "becomes actually involved or interferes with the Design-Builder's ability independently to design the project" (Loulakis and Shean 1996).

From the perspective of some Design-Builders, the Owner's involvement in the design development process may result in unnecessary delays, increase design and construction costs, frustrate their contractual and commercial expectations in the design development process, create conflict or inconsistency with other aspects of the design, negatively impact their working relationship with their own Engineering Consultant and/or trade Contractors, and frustrate design innovation opportunities.

In circumstances in which an Owner furnishes preliminary design and/or performance specifications that are implicated or involved in a deficiency in the Design-Builder's final design, disputes are likely to develop between the Owner and Design-Builder as to responsibility for such a deficiency. These disputes may involve a number of issues, such as:

- Was the deficiency due to some inadequacy in the development or definition of the preliminary design and/or performance specification?

- Did the Owner inappropriately intrude upon the Design-Builder's design development and finalization process, or change the design criteria or performance standards required of the Design-Builder during that process?

- Did the Owner assume responsibility for defective design due to the scope and nature of its role in the review and approval of the Design-Builder's design development submissions?
- Did the Design-Builder fail to develop the design in accordance with the preliminary design?
- Is the deficiency due to the Design-Builder's failure to detect errors or omissions, or the lack of coordination or other problems during the design-development process?
- Did the Design-Builder perform adequate constructability reviews during design development?
- Did the Design-Builder's failure to construct in accordance with the approved final design cause or contribute to the deficiency?

Causes of Defective Final Design. Defects in the final design may result from several causes, including the following:

- Deficient or inadequately developed or defined design criteria in the Owner-furnished preliminary design or performance specifications;
- Errors, omissions, or other inadequacies or deficiencies in the Design-Builder's development and finalization of the design;
- Failure of the Design-Builder to coordinate the design development performance of its multiple prime design professional consultants, all of which are under direct contract with the Design-Builder; and
- Failure of the Design-Builder to construct the project in accordance with approved design and/or contract documents.

The economic consequences of the Design-Builder's assumption of risk for deficiencies in the final design include

- Cost of revising design (redesign) to correct deficiencies;
- Direct cost of replacing incorrectly performed work-in-place with corrective work;
- Indirect costs (delay, disruption, acceleration) incurred as a result of design deficiencies;
- Claims of trade subcontractors and others who are impacted by or perform extra work or services due to design deficiency;
- Liquidated or other damages paid to Owner for failure to meet project milestones due to design deficiencies; and
- Loss of bonus fee due to late project completion resulting from design deficiency.

For the reasons previously discussed, contractual efforts to differentially allocate responsibility to the: (1) Owner for preliminary design; and (2) to the Design-Builder for final design may be neither effective nor feasible in the Design-Build method. The nature and extent of those relative and respective responsibilities typically will, and should, be defined in the Owner–Designer-Builder agreement. In some contractual approaches, the Owner will require that the Design-Builder warrant the adequacy and completeness of the design. In practical and legal effect, this warranty obligation requires that the design be perfect and that all risk of any design inadequacy (error or omission) be assumed by the Design-Builder.

The Design-Builder's design adequacy risk is further heightened by application of the so-called *Spearin* or *implied warranty* principle. The Spearin principle (which takes its name from a United States Supreme Court decision issued in 1918) has long been recognized as meaning that an Owner who issues design and construction documents to a Contractor impliedly warrants (subject to any overriding contractual limitations, disclaimers, or other qualifications) that the design and documents are adequate, complete, and accurate in all respects, and suitable for construction and performance of the project in accordance with the requirements of those documents. The Spearin principle applies to a Design-Builder who issues design and construction documents to its trade Contractors. Thus, in addition to having complete (typically warranty) responsibility to the Owner for adequacy of the final design, the Design-Builder also has implied warranty responsibility to its trade Contractors. This design adequacy risk typically is not insurable (at least to the extent that it requires performance beyond the professional standard of care, and may not be transferable to the design professionals retained by the Design-Builder, through indemnification or other contractual provision).

Consideration of the insurability of professional liability risk for defective design represents a significant threshold project management activity on major subsurface projects. Owner decisions regarding risk allocation for defective design need to be made in conjunction with project professional liability insurance coverages. Often, project-specific professional liability insurance policies—under which all professionals are insured under a single professional liability policy—are often the most effective way to coordinate and integrate coverage for professional liability risk exposure on a Design-Build project.

CONTRACTOR-LED DESIGN-BUILD: PROFESSIONAL PRACTICE AND RISK MANAGEMENT OBSERVATIONS

Clearly, as discussed, the role of an Engineer serving as a consultant to a Design-Builder is radically different from that same Engineer serving as a consultant to an Owner. Following are five topics that specifically address the professional liability implications of an Engineer serving as a consultant to a Design-Builder.

Potential Impact on Existing Client/Project Relationship Considerations

In any Contractor-led Design-Build approaches, the Engineer is engaged (directly or indirectly) by the Design-Builder (typically a Contractor). In these circumstances, the Engineer's client or contracting partner is the Design-Builder—*not* the project Owner. While it is accurate that Engineers—no matter by whom retained—owe certain basic legal obligations to the public (e.g., health, safety, or welfare), compliance with those minimal legal obligations is a far cry from establishing that the Engineer owes any degree of allegiance to the project Owner (with whom the Engineer has no direct contractual relationship).

In circumstances in which the Engineer has had direct contractual relationships with a project Owner in the Design-Bid-Build approach and enters into a subconsultant relationship with a Design-Builder on a Design-Build project for that same Owner, the project Owner may have an expectation that the Engineer will serve as its representative in the Design-Build process and owe allegiance to the Owner. The disappointment of those expectations may lead to friction and claims among the Owner, Design-Builder, and the Engineer and, worse yet, may disrupt or terminate longstanding professional relationships between the Engineer and the project Owner.

One commentator has described some of the tensions between the Owner and Engineer inherent in Contractor-led Design-Build as follows:

> [This raises] a whole host of legal, professional, and relational realignments for design professionals. As the term implies, most architects and engineers view themselves as

"design professionals" rather than as "subcontractors" working for a general contractor. Their professional practice, guidelines, ethical standards, and other traditions are based on the premise that they work as agents directly for owners. This becomes a continuing challenge for the design-build method of doing business and is a strain in the relationship between the builder and the designer that the owner should keep in mind. The reality of this tension may affect not only the contracting relationships, but also the outcome of the project.

When a construction contractor has the direct design-build contract with the owner, the contractor takes on all of the responsibilities, risks, and rewards of design. This is also a major realignment for the many contractors who do not price the risk of design errors, omissions, and design-related delay in their construction costs. These types of problems have been the source of numerous claims by contractors against owners under the design-bid-build project delivery system. Many contractors may not appreciate the professional judgment, time, and effort it takes for a designer to work with an owner to obtain approval for a design concept, design details and program elements. There is much handholding and give-and-take between the designer and the owner before an acceptable design is developed. In this form of design-build, the developmental aspects of the design process are generally minimized. This, too, may lead to tension between the design-builder and the owner, especially when the owner is unsophisticated. (Bramble 1999)

Many Engineers who have served as subconsultants for Design-Builders have expressed the frustrations and concerns associated with this "disappointed expectation/conflict" predicament. Although one may debate whether the circumstances described constitute a legal or ethical conflict of interest for the Engineer, it is clear that the Engineer's involvement as a subconsultant on a Design-Build project raises the potential for such tensions and disappointed expectations, which should be carefully evaluated and balanced, especially against a broader context of the Engineer and other, more traditional professional relationships with the same project Owner.

Professional Essence of Design-Builder/Engineer Relationship

The Engineer in the Contractor-led approach is not simply another subcontractor or vendor to the Design-Builder. In this role, it is critically important for both the Design-Builder and the Engineer to understand the distinction between providing professional services (involving the exercise of professional skill, acumen, experience, and judgment) in support of the Design-Builder and the more traditional trade or subcontractor, which provides labor or materials. As noted, Engineers are registered professionals who owe legal obligations to the public arising from the discharge of their professional services. In addition, the Engineer's services typically, and in most cases, require and depend upon the exercise of independent professional judgment and skill.

Some Engineers who have served in the subconsultant role on Design-Build projects have expressed frustrations and concerns about the Design-Builder's attempts to direct the design development effort or inappropriately control or intrude into the Engineer's exercise of independent professional judgment. Along these lines, it has been stated:

The designer's lack of independence in design-build may also raise professional and ethical concerns for the design profession. Problems may arise if a contractor controls the leadership of the design-build entity and puts pressure on the designer to focus on construction costs and possibly downplay the owner needs, material quality, or long term maintenance considerations. More critical are pressures that may have an effect on safety considerations. Some design-build contracts contain provisions requiring the design-builder to perform independent agent-type functions, such as inspections, certification of payments and certifications of substantial and final completion. Such clauses may be an inappropriate hybrid of design-build and design-bid-build agreements. The architectural and engineering communities also have raised related concerns. The lack of an independent and direct relationship between the designer and the owner is one such issue. Designers may resent being treated like "'subcontractors" rather than professionals. The concept of design professional practice developed over the years primarily in the context of design-bid-build contracting. Thus, the professional concept often involves qualification selection, rather than low price, and direct contact with the client. More than finances may be at stake. (Bramble 1999, Sec. 1.05, 25–26)

Clearly, a balance needs to be struck between the Engineer's ability to develop design within the constraints imposed by design criteria/performance standards furnished by the Owner (or otherwise established in the Design-Build agreement) and the appropriate level of independent judgment required to be exercised by the Engineer in that design development process.

Scope of Service and Communications Considerations

The Design-Builder, even more than many Owners in the traditional Design-Bid-Build method, sometimes unreasonably restricts the Engineer's level or scope of effort during the design development phase and may even allow for only on-call services of the Engineer during the construction process. This is unfortunate and, as is well known and understood, such unduly restrictive limitations on the Engineer's services (especially in subsurface projects) significantly increase the potential liability exposure for the Engineer.

These types of limitations are problematic, not only for the Engineer, but also for the Owner and its Engineering Consultants involved in Design-Build projects. More specifically, the design development process in the Design-Build method is both iterative and interactive and requires an appropriate level of access and communication between the Owner, its Engineering Consultants, the Design-Builder, and its Engineers. If the latter have not been retained by the Design-Builder to provide the appropriate level of professional support, the design development process will suffer in terms of timeliness, communication, and quality of the completed design. These problems are further exacerbated when the Design-Builder's Engineer has no role during construction or an unacceptably minimal or observational role.

An innovative approach to concern about scope limitation is to prescribe in the Design-Build RFP a minimal scope or level of effort required of the Engineer that is part of the Design-Build team and to provide for the ability of the Owner and its Engineering Consultants to communicate directly with the Engineer, provided that the Design-Builder is afforded an opportunity to be present and/or otherwise participate. This approach is unconventional in the sense that subcontractors of a general Contractor typically are not allowed to directly communicate with the project Owner or anyone other than the Contractor; however, Design-Build is an entirely different delivery approach and, as noted, Engineers are not just simply another subcontractor. For Design-Build to succeed and improve as a delivery method, we need to think out of the box and not be retarded by traditional limitations or conventional approaches.

Influence of Constructability and Cost Considerations in Design Process

By integrating the design and construction expertise into a single Design-Build team, the Design-Build delivery method affords an opportunity for the design process to more directly and contemporaneously benefit from the constructability and cost estimating experience and expertise of the Contractor member of the team. This integration has substantial benefit on subsurface projects, given the interrelationship between design adequacy, the construction means and methods and equipment to be employed in the construction of the project, and the behavior of subsurface conditions. The Contractor's input during design development relates to such matters as costing and estimating; value engineering; analysis of site, geotechnical, environmental, and other site-specific information for the purpose of project planning and sequencing; constructability review; preliminary scheduling; design review for errors or omissions; long-lead item procurement; arranging for trade Contractor involvement; and identifying design delegation opportunities. At the same time, for many Engineers, interacting with a Contractor during design development—especially a Contractor to whom the Engineer is contractually subordinate—is a somewhat different experience. Simply put, it is inaccurate to say that the service activities for the Engineer in this role essentially remain the same as in Design-Bid-Build and the only thing that has changed is the identity of the client. However, the client's identity —a Contractor rather than an Owner—has a dramatic impact on the nature and emphasis of the Engineer's service effort. Unquestionably, the Engineer must be receptive to the interaction and input of the Contractor during the design development process with respect not only to the design of permanent work, but also to constructability, means and methods, and equipment selection considerations. As stated in a draft revision to American Society of Civil Engineers Manual 73, entitled "Project Delivery Systems":

> A key point to be noted in the Design-Build project is that the Design Professional Party to the Design-Build organization worked for the Contractor/Sponsor. As such, the attention of the Designer and allied professionals must be appropriately focused on the requirements and expectations of the Builder. Issues such as constructability, use of particular equipment or erection methods, and the choice of construction materials may control design and be directed by the Contractor. While the resulting designs and facilities must meet the Standards and state of requirements of the Owner, beyond that, the design focus must be on the preferred means and methods of the Contractor as the client. This is a very different role and focus for most designers and can be expected to require some deliberate cultural adjustment.

Another author has commented on the impact of cost considerations on the design professional:

> Another outcome of soliciting design input from construction players is a focus on cost reductions. A bottom-line perspective is often obtained because a Design-Build entity usually has a price limitation, if not a fixed price, agreed to with the Owner. This cost orientation may also permeate the relationships with trade subcontractors who perform portions of the construction work on a fixed price basis in an attempt to achieve savings. Tension may arise when the designer of record has to finalize construction requirements in conformity with engineering standards, safety requirements, building codes, esthetic considerations, and other professional considerations. If the constructor plays the lead role in the Design-Build process, it may increase the emphasis on bottom-line considerations at the expense of engineering or operational considerations. This entails more responsibility and risk for the constructor but increased pressure for the designers.

Similarly, if the scope or price of the design services are strictly limited, tension may also arise within the Design-Build team because of the professional standards conflict, creating a source of additional risk. (Bramble 1999)

Contractors leading the Design-Build team and their Engineering Consultants need to better appreciate each other's risks. As Essex and Zelenko (2000) have stated:

> In the relationship between the contractor and design engineer, there is room for greater sensitivity to the financial and professional risks that each entity faces. While the designer is working within a budget, the greater financial risks will be shouldered by the Contractor—this must be respected by the engineer, in the time that it has taken to develop design changes "on the fly" during construction. The need to have effective design decisions made quickly can be better anticipated by having adequate budget to fund the necessary professionals on site if necessary to meet those challenges. Also, sophisticated analyses may consume more time (cost) in their development than is saved through the more efficient design.
>
> On another front, the design engineer must uphold a professional standard in carrying out design decisions and certifications. Economizing a design to reduce construction costs creates a potential dilemma for the designer—the professional code that forms the basis for engineering licensing follows the design engineer's judgments and actions well beyond the construction "warranty period" of any particular project. Contractors must respect this role and responsibility that the designer carries as Engineer of Record.

Professional Liability Considerations

There are several important professional liability implications for Engineers serving as subconsultants to a Contractor in Contractor-led Design-Build.

1. The manner and fairness in which risk is allocated between the Owner and Design-Builder will impact allocation of risk between the Design-Builder and its Engineer subconsultant. As a general proposition, if risk is unfairly or inappropriately allocated in the prime agreement, it is probable that the Design-Build Engineer will—by virtue of *flow down* or *incorporation by reference* provisions—be subjected to a similarly unfair risk allocation burden. When that occurs, the probability of professional liability exposure is significantly increased for the Engineer (Hatem 1998b, 313–340; Hatem 1999).

   Examples include

   - Warranty/guaranty provisions with respect to design adequacy;
   - Complete risk transfer to the Design-Builder for unanticipated site and environmental conditions;
   - Redesign at no cost due to building code or regulatory changes that impact design; and
   - Responsibility for obtaining all local, state, or federal or other governmental approval required for the design and construction of the project.

   In addition, Contractors (and their counsel) may not be sufficiently sensitive to and/or conversant with liability and risk management concerns from the standpoint of the Engineer and thus not adequately represent the Engineer's interest. For those reasons, it is important that the Engineer (and its counsel) have a place at the table in the negotiation of the prime Design-Build agreement—at least with respect to issues that pertain to the Engineer.

2. In addition to risk allocation provisions in the prime agreement, other important factors may impact the professional liability exposure for Engineers in a Contractor-led Design-Build project:

    a. The range of services to be provided by the Engineer is likely to be more encompassing than the Engineer's scope on a traditional Design-Bid-Build project, particularly as relates to retention of subconsultant disciplines. Specifically, consistent with the Design-Build hallmark principle of single-point responsibility, it is probable that the Engineer subconsultant will be required to retain all professional service disciplines, including geotechnical, environmental, asbestos, hazardous waste, civil, site, survey, permitting, and other services. On the even more nontraditional end of the spectrum, professional services such as cost estimating, accounting, public relations, and legal services may be required to be retained by the Engineer. The greater the range of subconsultant retention, the greater the range of potential professional liability exposure, especially if the service activities fall beyond those traditionally retained and managed by the Engineer.

    b. At the other end of the spectrum, the structural relationships for the Design-Build project may involve the retention of multiple, prime Engineers by the Design-Builder or the retention of one prime Engineer and multiple Engineers by specialty trade Contractors. In these circumstances, liability exposure issues will arise out of design coordination responsibilities and the potential ambiguity, fragmentation, and diffusion of responsibility for final design among multiple Engineers (Hatem 1998c, 1–21).

    c. In many significant Design-Build projects, the Owner will correctly perceive the need and/or desirability of retaining its own Engineering Consultant(s) to develop conceptual or preliminary design and represent its interests in the Design-Build procurement process; to review design development submissions by the Design-Build team; to review submittals, certifications, payment and requisitions; and to observe construction during the construction phase (Hatem 1996, 8–20). The degree of involvement of the Owner in design development and review, as well as the expertise and degree of discretion delegated to the Design-Builder (and its Engineer subconsultant) to develop the design, are factors that will significantly affect professional liability risk for the Owner and its Engineering Consultant, as well the Engineer who is a subconsultant to the Design-Builder (Hatem 1998c, 1–21; Hatem 1996, 8–20). In this regard, one commentator has stated:

        Risk allocation in design responsibility is a key difference between design-build and the traditional design-bid-build delivery system. Traditionally, there has been a clear demarcation between design responsibilities and construction responsibilities. But the design-builder has increased responsibility for the defects and deficiencies in design, particularly those aspects of design that it prepared. Not all of the design aspects are performed by the design-builder. For example, prior to selecting the design-builder, the owner may retain a designer to develop a project concept, a program, and criteria that may become the project requirements handed over to the design-builder to implement. The owner has responsibility for these aspects of design. There is no bright-line test that separates this aspect of design from the more detailed engineering and design performed by the design-builder. However, design responsibility is shifted from the owner to a greater extent under design-build than in more traditional project contracting methods. (Bramble 1999)

    d. The greater the involvement of the Engineer in the design and/or implementation of construction means and methods or safety precautions or programs, the greater the Engineer's liability exposure for bodily injury and property damage claims and exposure under OSHA (Hatem 1996, 8–20; Hatem 1998d, 1–16).

3. In Design-Bid-Build, the Contractor typically is furnished with a complete design by the Owner, which carries with it an implied warranty that the Owner-furnished design is complete and accurate in all respects, and constructable. If that warranty obligation is breached, the Contractor (subject to the terms of the contract documents) may be entitled to an appropriate equitable adjustment from the Owner. In contrast, in Design-Build, the Owner provides the Design-Builder with—at most—a conceptual or preliminary design and often disclaims the completeness, accuracy, and/or constructability of that design. The Design-Build RFP typically will explicitly state that the Design-Builder will be responsible for the development, finalization, and accuracy of the design, and for the adequacy, completeness, or certain performance characteristics of that design. As such, in a Design-Build project, it is unlikely that, in most instances, the Design-Builder will have any significant recourse against the project Owner for defective design. As a practical matter, however, this lack of recourse against the Owner translates into a higher risk of errors and omissions claims by the Design-Builder against its Engineer subconsultant.

4. In Design-Bid-Build, it is relatively rare for the Owner to assert a claim against the Engineer due to delays or other consequential impacts of errors/omissions during the design development phase; adverse economic impacts of defective design typically manifest during the construction phase of the project and, consequently, claims against the Engineers are asserted then or after project completion. In contrast, in Design-Build, the Engineer's errors/omissions may well surface much sooner in the cycle. For example, if the Engineer has erred in providing preliminary design or related information to the Design-Builder which, in turn, has relied upon in pricing the project, the adverse impact may be experienced by the Design-Builder sooner than the commencement of construction.

SUMMARY

The Design-Build method provides an opportunity to develop and enhance the improved contracting practices utilized for subsurface condition risk allocation and fair dispute resolution in the Design-Bid-Build method. The collaborative effort and interaction between the Owner and its Engineering Consultants, and the Design-Build team, prior to the point of contract formation, provides an opportunity to enhance the design and construction process and to more clearly and objectively define reasonable and fair expectations and risk allocations. In this way, also, risk-transfer triggers can be established for subsurface conditions that vary from the contractually defined assumptions or baselines. In the opinion of this author, given the opportunities afforded by the Design-Build method for substantial improvements in the subsurface condition risk allocation process, it would be a serious mistake to disregard the lessons learned and improved contracting practices developed in the Design-Bid-Build process and revert to some of the more problematic practices, such as inclusion of unqualified disclaimers and exclusion of DSC clauses.

The successful delivery of a major subsurface project—a delivery that is on schedule and within budget—requires a close and dynamic interaction between professionals who are responsible for

- Investigation, evaluation, and reporting of subsurface geotechnical and environmental conditions;
- Conceptual/preliminary design of the project;
- Final design; and
- Management of project construction.

In addition, in a Design-Build subsurface project, the Owner's Engineering Consultant, which is responsible for preliminary design, must interact with the Design-Builder in a manner that will not create ambiguity or confusion over contractual allocation of risk between the Owner and the Design-Builder. These close and dynamic interactions involve the performance of a wide range of professional services and obligations, not only of the Owner's Engineering Consultant, but also the Engineers who are part of the Design-Build team. In addition, these interactions pose substantial effective contractual risk allocation and liability insurance challenges.

Accomplishing the successful delivery of a major subsurface project when various project participants have independent contracts and overlapping roles and relationships presents substantial challenges, not only in terms of clear and effective risk allocation, but also in terms of the effective and cost-efficient insurability of that risk and the prompt and cost-effective resolution of any disputes and claims involving the performance of professional engineering services. The inability to clearly ascertain responsibility among the various engineering firms has a negative impact on relationships among project participants and, invariably, negatively impacts the ability to successfully deliver the project on schedule and within budget.

These challenges are present on most subsurface Design-Build projects. Although final design responsibility will be centered on the Design-Builder, the nature of Design-Build subsurface projects requires that the design be

- Based upon conceptual or preliminary design developed by the Owner's Engineering Consultant;
- Based upon an appropriately defined and managed geotechnical and environmental investigatory and evaluative program;
- Periodically evaluated by the Owner and its Engineering Consultant during design development for cost, scheduling, and constructability considerations; and
- Regularly reevaluated during the construction process for potential design revisions due, for example, to value engineering proposals, the Design-Builder's selected means and methods, and/or ground or field conditions encountered during the performance of the work.

Risk allocation may be significantly impacted by Owner choices with respect to the various options for assignment of roles and responsibilities among project participants, such as which is responsible for subsurface investigation and the degree and extent of the Owner's design development; and the discretion or latitude given to the Design-Builder in the design development and finalization process. Important decisions by the Owner on these and related topics will materially impact risk allocation among project participants.

Despite the significant and, to some degree, potentially nontransferable (or insurable) design adequacy risk assumed by the Design-Builder, there are some factors to consider in the Design-Build context that may have the effect of mitigating that risk. Some statistical information presently available from the professional liability insurance industry supports the conclusion that the frequency of claims against design professionals in the Design-Build method is less than in Design-Bid-Build. There may be many reasons for this reduced frequency, one of which may be improved and enhanced quality relationships (i.e., less adversity) among Design-Build team members, which reduces the number of claims against the design professional members of the team. Another explanation may be that a close and collaborative relationship between the Contractor and Engineer, especially during design development, significantly reduces the opportunity for errors or omissions in the final design or constructability problems surfacing during construction.

The contractual approaches accepted by the Owner and Contractor at the point of contract formation should be respected, and the advisability or fairness of those risk allocation choices should not be second-guessed or rewritten by others in the light of actual performance issues or problems that arise during construction (Sweet 2000).

This principle is especially important in the context of publicly procured construction contracts in which multiple Contractors competitively bid on the same terms and conditions and work scope (that typically include risk allocation provisions for subsurface conditions). Under such circumstances, it generally would subvert, undermine, and be disruptive of the competitive bid process, as well as unfair, were the successful bidder subsequently relieved of the consequences of those risk allocation provisions that it bid upon and accepted as a part of its contractual undertaking. In addition, were that to occur, the Owner would be deprived of the benefit of competitive pricing resulting from the subsequently revisited or retroactively modified risk allocation provisions.

As in the case of subsurface conditions, decision-making in the Design-Build method would benefit from conscientious consideration, evaluation, and adaptation of principles that have traditionally applied and worked reasonably well in the Design-Bid-Build context. The guiding principles in this process should be fairness and clarity, along with the recognition that the hallmark Design-Build principle of single-point responsibility does not necessarily mean or require complete and absolute risk transfer under all circumstances to the Design-Builder. In addition, the project participant in the optimum position to control a particular risk of defective design should be clearly allocated the responsibility for that risk. In order for these principles to be successfully and fairly implemented in the Design-Build context, the Owner must understand that the greater its (or its independent Engineering Consultants') control, decision-making, and dominance in the design development and finalization process, the less likely it will be able to effectively transfer all, or perhaps even most, defective design risk to the Design-Builder. Finally, Owner decision-making regarding defective design risk allocation should always be undertaken in a context of understanding the consequences of those decisions in relation to the insurability of that risk.

BIBLIOGRAPHY

American Society of Civil Engineers. "Project Delivery Systems." Revision to Manual 73.

Bramble, B.B. 1999. Design-Build. In *Design-Build Contracting Claims*. Edited by B.B. Bramble and Joseph D. West. Aspen, sec. 1.03.

Cohen, D. 1995. Minimizing conflicts between design professionals and contractors. *Construction Briefings*, 2nd Series. Washington, DC: Federal Publications, October.

Essex, R.J. 1997. *Geotechnical Baseline Reports for Underground Construction—Guidelines and Practices,* Technical Committee on Geotechnical Reports of the Underground Technology Research Council. New York: American Society of Civil Engineers.

Essex, R.J., and Zelenko, B. 1999. Design/build—the pros and cons. *Tunnel Business Magazine.* June.

———. 2000. Design/build for underground construction. Presented at North American Tunneling Conference, Boston, MA, June.

Hatem, D.J. 1996. Design-build: professional liability and risk management issues for design professionals. *The CA/T Professional Liability Reporter* 2, no. 2 (December).

———. 1997. Geotechnical baselines: professional liability implications. *The CA/T Professional Liability Reporter* 3, no. 1 (October).

———. 1998a. Professional liability and risk allocation/management considerations for design and construction management professionals involved in subsurface projects. In *Subsurface Conditions.* Edited by D.J. Hatem. New York: John Wiley & Sons.

———. 1998b. Design delegation: risk management/allocation considerations for design professionals. *The CA/T Professional Liability Reporter* 4, no. 1 (November).

———. 1998c. Changing roles of design professionals and constructors: risk allocation, management and insurance challenges. *The CA/T Professional Liability Reporter* 3, no. 3 (May).

———. 1998d. *Subsurface Conditions: Risk Management for Design and Construction Management Professionals.* New York: John Wiley & Sons, pp. 313–340.

———. 1999. Risk allocation for subsurface conditions: design-build projects. *The CA/T Professional Liability Reporter* 4, no. 3 (July), 1–17.

Hinkle, B. 2001. Does the proliferation of ADR hinder the development of construction law? *The Construction Lawyer* (Spring), 4.

Klein v. Catalano. 1982. 437 N.E. 2d 514, 525 (Mass.).

Loulakis, M.C. 2001. Design-build liability issues. In *Design-Build Contracting Handbook,* 2nd ed. Edited by R.F. Cushman and M.C. Loulakis. Aspen, sec. 1.08, pp. 25–31.

Loulakis, M.L., and Shean, O.J. 1996. Risk transference in design-build contracting. *Construction Briefings,* 2nd Series. Washington DC: Federal Publications, April.

M.A. Mortenson Co. Armed Services Board of Contract Appeals (ASBCA) 39978, 93-3 BCA par. 62.

Pitt-Des Moines, Inc. ASBCA 42838, 96-1 BCA 27941.

Sweet, J. 2000. *Legal Aspects of Architecture, Engineering and the Construction Process,* 6th ed., Chapter 25. Pacific Grove, CA: Brooks/Cole Publishing.

CHAPTER 4

Team Structures and Relationships

Thomas F. Peyton, P.E.
Parsons Brinckerhoff, New York, N.Y.

John A. Harrison, P.E.
Parsons Brinckerhoff, Sacramento, Calif.

INTRODUCTION

This chapter looks at the Design-Build method as it relates to the structure of teaming arrangements and relationships among the project participants. Design-Build differs from the traditional Design-Bid-Build method in both the business and legal roles of the participants. This requires a re-education about who is responsible to whom, who makes the decisions, and who communicates with whom. Although these nontraditional roles are sometimes not easy to accept, a clear understanding of them is essential to a successful Design-Build project. Failure to understand and work within these new roles can lead to friction among the parties, disputes, and increased costs.

In the seven years since this book was originally published, the industry has seen a marked increase in the use of public-private partnerships (PPPs) as a vehicle to deliver needed infrastructure projects in an environment where public infrastructure and services needs far exceed the capacity of government funds to provide these projects and services. The PPP adds additional players into the Design-Build equation, including sometimes an operation and maintenance company and bank lender(s).

In most projects involving subsurface work, the players include

- Owner,
- Design-Builder,
- Owner's Engineering Consultant,
- Surety,
- Insurance company(ies), and
- Public and regulatory agencies.

ROLES OF TEAM MEMBERS

Each player has a unique role in the Design-Build arena (Figure 4.1). All parties must understand the needs and expectations of the others, and there must be clear communication among the parties. These roles may be different in many details from a traditional Design-Bid-Build approach, and these differences must also be understood to allow for a mindset shift to Design-Build.

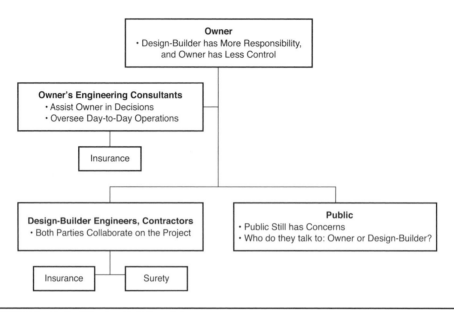

FIGURE 4.1   Typical Design-Build participants

Owner

The Owner typically is responsible for developing and providing a clear definition of the project and for communicating that definition to all parties on the Design-Build team. The Owner should also make known its expectations for the Design-Build process and from each of the parties. The project should proceed only after there is a clear project definition, including requirements and performance specifications, which are communicated to and understood by the remaining team members. The Owner should also be responsible for providing all available information about the subsurface conditions. The Owner should provide a clear, complete request for proposals (RFP) containing the definition of the project and any other requirements. The RFP should include a clear process for the selection of the successful Design-Builder. In cases where insufficient subsurface information is available at the time of the RFP to accurately estimate the cost of construction, contract provisions should be included to share the risk between the Owner and the selected Design-Builder for reasonable cost adjustments once this information can be obtained. This should minimize the need for unnecessarily large contingencies in the proposal price.

The Owner must be concerned with building public support for the project to help ensure that the project is not delayed or disrupted by opposition groups. The National Environmental Policy Act and study requirements such as the Major Investment Study have empowered the public to have a voice in the planning of public projects, and it wants and expects to be involved. Owners of public and private projects should recognize and cultivate this public involvement as a way to expedite their projects.

Design-Builder

The Design-Builder role combines the design and construction of the project. The Design-Builder is the single point of responsibility for understanding and accepting the Owner's requirements, completing a design that satisfies these requirements, and then constructing the project. The

Design-Builder is responsible and liable for the design problems as well as any construction delays and defects. It is vital that the Design-Build team develop a working relationship among team members that is supportive rather than adversarial.

Contractor

The Contractor's role typically includes providing the material, equipment, labor, and management to construct the project in compliance with the plans and specifications and working with the Designer to ensure that the design is buildable and maintainable. Both have a stake in working together to maximize each others' profits, to control costs, and to minimize disputes and design errors that could impact the success of the project.

Designer

The Designer's typical role is to understand the Owner's vision of the project and to provide a design that satisfies all of the project's known short- and long-term requirements. The Designer is responsible to the Contractor to produce a buildable design that meets the Owner's requirements. The Designer has a duty to the public to produce a safe design. In most states, this duty is a statutory requirement of the profession and is mandated by the licensing laws by which the professional practices. The respective success of the Designer and the Contractor are linked. Each must do what it can to solve problems during construction, so these problems do not adversely affect project completion.

Engineering Consultant

The Owner may retain an Engineering Consultant for several reasons. The Owner may not have sufficient staff to prepare the project requirements and administer the project by itself. There may be legislative requirements for oversight of, and assistance to, the process. For whatever reason the Owner chooses to employ the services of a professional firm, the roles and responsibilities must be clearly stated and understood. The role may be to assist the Owner in putting together the design definition and criteria, to help prepare the RFPs, or to help evaluate the proposals and to assist in selecting the Design-Builder. In some cases the Owner's Engineering Consultant will continue to assist the Owner by providing technical assistance or dispute and claim assistance throughout the life of the project. Some Design-Builders will resent having to deal with an additional party other than the Owner. This reluctance must be recognized, and procedures must be put in place to satisfy the requirements of the project.

Surety

The surety is a company that accepts risks transferred by parties in a construction contract to a third party (i.e., the surety). This risk transference is usually accomplished by the Owner requiring the Design-Builder to provide a performance bond. The surety promises that in the event the Design-Builder defaults on its obligation to complete the project, the surety will step in and fulfill the Design-Builder's obligation. The surety may accomplish this by providing another Contractor to complete the work, by paying for the Owner to obtain another Design-Builder, or by providing funds and oversight to the original Design-Builder so it can complete the work.

Insurance Companies

Insurance companies provide a means to manage risk. Standard insurance products have been developed to provide coverage for many of the risks typically associated with the design and construction of a project. These products also have conditions, exclusions, and qualifications that must be understood in order to seamlessly provide the risk coverage and management that each of the parties desires for the project.

Public

The public has a role in all projects because it is usually either the end user or the affected party during the facility's construction and operation. Widespread acceptance of public rights and obligations in the project process can affect what, how, and even if a project is constructed. All parties must understand that the public is not to be ignored during the planning and construction process.

DESIGN-BID-BUILD COMPARED TO DESIGN-BUILD

In a traditional Design-Bid-Build approach (Figure 4.2), the Owner contracts with an engineering firm to design the project. Considerable time is usually available, so that the Owner can explain the general outlines or requirements of the project to the Engineer. Together they can develop alternatives and arrive at an optimum solution or design. The Owner does not need to have a well-defined scope of the project at the beginning because it can develop one with its Engineer. As more data become available, necessary changes can be made before the project is advertised for construction bids. The Owner has the advantage of having the Engineer working directly for it, which obligates the Engineer to prepare the design both to support the requirements of the project and to protect the general public, which will be the end user. The Engineer's design does not have to be perfect or free from defects but only satisfy a standard of care that a typical Engineer would meet. The Owner retains the right to have a great deal of input as to what goes into the design and to change the design as new circumstances are uncovered. The Owner does, however, warrant to the Contractor that will ultimately build the project that the design is buildable and free of defects.

One of the Owner's most important roles is dealing with the public, which must have a sense that its concerns and desires will be heard and incorporated into the project procedures. So the Owner can develop and maintain public support, this effort to develop a relationship of trust usually begins early in the planning stages and continues throughout the project.

Once the plans and specifications are completed, the Owner, with assistance from the Engineer, invites Contractors to prepare a bid or tender for the work as defined by these documents. The Owner represents to all the bidders that the work can be constructed as shown and invites and requires these bidders to tell the Owner if they find errors in the documents that would cause the bidders to think otherwise.

The next step is for the Owner to contract with a Contractor to build the work. The Contractor is then responsible for constructing the work in accordance with the Owner-supplied plans and specifications. The Contractor has the right to assume that the project is buildable as designed. The Contractor only warrants work performed and the equipment or material supplied and does not warrant that the project will perform as designed.

The Owner may contract with a firm to manage the construction. This Construction Management firm assists the Owner in evaluating construction bids, recommending the firm that is the low responsible bidder and conducting the preconstruction meeting to ensure that the Contractor understands the requirements of the contract and what will be expected from it. The Construction Manager then administers the day-to-day operations of the work, processes progress payments, and

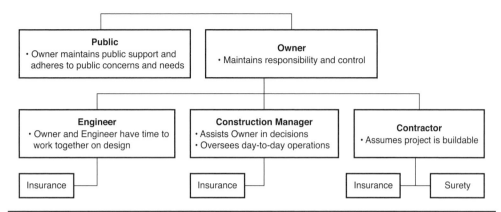

FIGURE 4.2   Typical Design-Bid-Build structure

assists the Owner in evaluating and processing change orders and claim requests from the Contractor. The Construction Manager may also maintain the project master schedule and perform quality assurance functions. In Design-Bid-Build, each participant has a contract with the Owner; thus, each role and the communication among the parties are understood.

In Design-Build, the Owner has the benefit of one party accepting full responsibility for the project, providing an early guaranteed price for the work as specified and quicker project delivery. In exchange for these benefits, the Owner surrenders some control over the product, the process, and unrestricted access to the traditional Owner-Engineer interplay during the work. The Design-Builder takes on a substantially different role with different responsibilities because of the additional liability for design adequacy, successful performance of the project, and the usual construction risks.

As was pointed out earlier, it is important for the Owner to have a clearly defined scope for the project, so there are no questions as to what is required. It is just as important for the members of the Design-Build team to have a written teaming agreement in place, early in their relationship, that confirms the responsibilities and liabilities of each member. This teaming agreement should specify who performs each role and the extent as well as the limits of their influence over the project. Deliverables should be listed along with their required schedule and remedies for missed deliverables. There must be clear understanding of the relative financial participation of each party, both as risk and reward. Dispute resolution mechanisms should be included that can be implemented if and when the parties do not agree. Each party would be wise to treat the Design-Build teaming arrangement like any other contract and have procedures in place for proper administration of the agreement.

As a minimum, the teaming agreement should address (Friedlander 1997)

- **Team structure.** What is the legal setup of the business relationship, and how much capital will each party put at risk or otherwise contribute? Which one controls the organization, and what are the lines of communication? Which one communicates to the Owner and under what circumstances? (A general discussion of some common structures follows later in the chapter.)

- **Risks and rewards structure.** A risk matrix should be prepared and each party assigned a lead role, a supporting role, or a shared role in dealing with the risks. How will the parties share in the profits or losses? When are the payments to be made?

- **Roles for each phase.** The agreement should then address each phase of the project and the roles and responsibilities of each of the parties. These phases should include a proposal, a design, and a construction phase.

- **Preparation costs.** During the proposal phase, the agreement should contain language about how each party will handle its costs for preparation efforts. In addition, the agreement should discuss how those costs will be recovered if the Owner provides a stipend for this aspect of proposal preparation.

- **Contractor and Designer roles.** During the design phase, what roles will the Contractor and Designer play? Which will own the design? Which will provide the cost estimates? How will issues of design authority be resolved in the event of a dispute between the Designer and the Contractor?

- **Unanticipated problems.** During the construction phase, how will the parties work together to solve unanticipated design, site, or subsurface problems?

- **Insurance.** The issue of the transfer of risk using insurance must be addressed, and a thorough review of existing and potential insurance products must be conducted, including indemnity provisions. Traditionally, Designer's errors and omissions insurance usually excludes coverage for defective construction work, and a Contractor's general liability coverage excludes any coverage for design services. These differences must be understood and coverage arranged to fill these gaps. This is why a review by outside insurance professionals is recommended. The review will also address the deductibles associated with these respective policies and how the resulting uninsured risk will be handled.

- **Dispute resolution.** A dispute resolution process must be established. Because the Design-Build team is composed of members who look at risk differently and approach construction from varying perspectives, a method of dealing with disputes is essential. All members of the team must be able to understand and to easily utilize the dispute resolution method. This process could include partnering, a dispute review board, mediation, or arbitration as means to address issues that arise before they become obstacles to the Design-Build process.

- **Cooperation.** How will the parties deal with each other for this project and future projects? Are the parties dealing exclusively with one another, or are they free to solicit other teaming arrangements? Can the parties compete against each other if they are, or are not, successful on this project? Do the parties agree to keep each others' information and work products confidential?

This list is not meant to be comprehensive and must be evaluated on a case-by-case basis. The agreement must capture the idea that it is in the best interest of all parties to work together to maximize each others' profit and to solve any design or construction issues that arise, so that a buildable design ultimately reflects the expertise of the Contractor and satisfies the Owner's requirements.

And finally, it is important to note that Design-Build does have its limitations. Some of the most common disadvantages cited for using this process include

- Because the selection criteria are subjective, they could be abused by manipulating the way the criteria are weighted. There is a belief that Design-Bid-Build (low bid) eliminates this potential problem.

- The ability of the Owner to control the design is diminished because the Designer is working for or with the Contractor and not for the Owner.

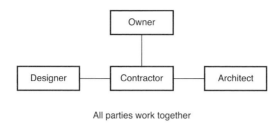

All parties work together

FIGURE 4.3 Contractor-led team

- In addition to not controlling the design, the Owner loses the traditional system of checks and balances inherent in the Design-Bid-Build system, in which the Engineer works for the Owner.

- The Owner may be forced to hire additional firms to act as its advisers, execute inspection services, or fulfill miscellaneous duties normally performed by its Engineer, thereby increasing the cost to the Owner and offsetting any potential savings expected from using Design-Build.

- This method may not be allowed under the current bidding or bid shopping laws of the state, city, or other entity.

CONTRACTOR-LED DESIGN-BUILD

The Contractor-led team is the form that has been used for the majority of Design-Build projects to date (Figure 4.3). A survey indicated that 57% of Design-Build projects in the United States are led by contractors, while Architects and Designers assumed the leadership role in only 17% and 15%, respectively (Smith 1996).

Contractors are better equipped to manage construction risk than Designers. In part, this is due to the relative values of the risk involved and access to control or management of these risks. Contractors are exposed to substantial monetary risks in every project they undertake and also have ultimate authority over construction means and methods. These observations are especially important in subsurface projects. When unanticipated problems arise, means and methods can frequently be revised to minimize adverse impacts. Designers are typically neither organized nor experienced to deal with the financial risks inherent in subsurface construction. While it can play an important supporting role, it is unusual for the Designer to take the lead role in performing subsurface construction. Most Designers would not have the bonding capacity needed for a large subsurface construction project; do not have as much experience estimating costs in enough detail to manage construction; and do not have the experience to deal with and manage craft unions. Although Designers may be exposed to substantial professional liability risk, the impact of this risk can frequently be reduced or managed through appropriate insurance and good management. At the same time, Designers would be ill-advised to undertake risks that materialize only during field operations because they would not have direct access to the management tools and authority required to control or otherwise mitigate such risks.

In the Contractor-led approach, the Designer works for and with the Contractor instead of for the Owner. This can reduce the frequently adversarial relationship that exists in Design-Bid-Build. If issues arise during construction, it is in the team's best interest to solve them quickly to move the project forward. The Contractor participates in writing and reviewing the plans and

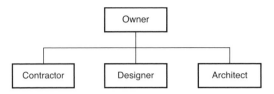

FIGURE 4.4   Designer-led team

specifications, which makes the Contractor more sensitive to constructability considerations and responsive to their anticipated means and methods.

Contractor-led teams have the financial strength (i.e., net worth and bonding capacity) as well as the risk-taking mindset to take on large Design-Build projects. However, many larger engineering firms are diversifying and setting up their own Design-Build organizations and bidding on Design-Build jobs at risk. This is commonly done either as joint venture partners with Contractors or as the lead firm of a Design-Build team. The advantage to the engineering firm to lead a Design-Build team or participate in a joint venture is the larger profit potential if the risks can be properly managed. A potential advantage to the Owner of having a strong Designer-led team is a greater level of technical expertise in the design and management of the project. The advantage of the more conventional Contractor-led approach is the emphasis on constructability and experience in managing subsurface construction.

DESIGNER-LED DESIGN-BUILD

Designer-led teams (Figure 4.4) are gaining popularity in the United States, but this approach places the Designer in unfamiliar risk territory. Designer-led Design-Build teaming arrangements can offer attractive project delivery options to the Owner. One such option is the *sequential Design-Build* method. In this arrangement, the Owner has the opportunity to try on the business relationship before committing to a full Design-Build construction effort. The Owner can start a project in much the same way as the traditional Design-Bid-Build by contracting for design services only. After the Owner is comfortable with the Designer or has secured the approval of its board, it can convert the project to Design-Build by contracting with the Designer or the Designer's Contractor to build the project (Friedlander 2000).

As with all forms of Design-Build teaming arrangements, these methods require that the Designer and Contractor avoid the adversarial relationships common to Design-Bid-Build projects and work together, so that each party successfully completes the project and makes money. In order to ensure that this cooperation occurs, each party should share in the profits and losses from the work. The Designer stands to earn significant revenues from sharing the Contractor's profits in addition to its traditional design fee. There is also the potential to share in the savings generated by the Contractor's material purchases and work sequences. It should be clearly understood that the Designer is unlikely to share disproportionately in the construction profits if it assumes few of the risks. The apportionment of risk and potential reward in the teaming agreement is basic to a fair balance. The Designer must be especially prudent in undertaking unfamiliar risk in order to share potential profits. One area in which increased profits may be available to the Designer is in sharing cost savings, proposed by the Contractor, that are incorporated into the design or acted upon during the construction phase.

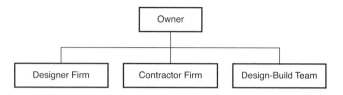

FIGURE 4.5 Contractor/Designer joint venture

When a project demands a highly sophisticated technical understanding of the design and construction challenges, because of either the difficulty or complexity of the subsurface work, Designer-led teams may offer the Owner the services or comfort level it seeks for the project. The Designer may employ a construction manager-at-risk approach to the work, in which, on the one hand, the Designer retains the professional relationship with the Owner and, on the other, sub-contracts the actual construction to a qualified Contractor, which in turn bonds and insures the construction portion of the project. In order to take the lead, the Designer must have sufficient financial resources, some bonding capacity, a suitable legal framework, and some construction experience. Few Designers can fill these requirements. This arrangement also follows a more traditional sequence, resulting in less perceived resistance to the unknown structure.

CONTRACTOR/DESIGNER JOINT VENTURE

This method is not very common because of the joint and several liabilities that are imposed on the parties (Figure 4.5). Joint and several liability means that each party is legally responsible for the acts or omissions (or other deficiencies) of the other. Although it is not unusual to have joint ventures among like parties, such as Contractors on a large project, a joint venture among dissimilar parties is difficult to manage because each approaches the work from a different perspective. It is unlikely that the relative values and risk of each member's contribution to the activity will be similar. There are examples of Architect/Engineer–Construction Manager contracts, however, notably the Superconducting Super Collider (SSC) project in Texas, which had extensive subsurface construction (72 miles of tunnels and 40 shafts), where a major engineering company and a construction company formed a joint venture to design and manage the construction.

As in construction manager-at-risk, the actual construction on the SSC was required to be performed by Contractors selected through a competitive bidding process, but that process was managed exclusively by the prime contractor and the construction works. Contractors held sub-contracts with the prime contractor, and each subcontractor brought its required bonding. All project participants were wrapped up under an Owner-controlled insurance program that provided comprehensive general liability, builder's risk, worker's compensation, personal liability/property damage, and professional liability coverage. In addition, contractual language limited the liability of the prime contractor with respect to project performance risks and losses but also subjected the prime contractor to cost performance risks on the entire program of work and services through an incentive performance fee. Such contracts are complex and represent unique solutions to meet the needs or desires of particular Owners. These arrangements can be effective as well. In the SSC case, the cost of the constructed work, when the project was terminated, was around $100 million less than the target cost set for those works.

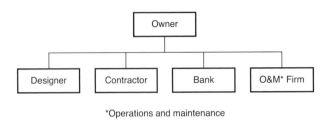

*Operations and maintenance

FIGURE 4.6   Public-private partnerships

PUBLIC-PRIVATE PARTNERSHIPS

A PPP contract is between a public sector authority in need of a specific project or structure and a private party capable of providing a team to produce the needed project or structure. In the PPP, the private party provides a public service or project and assumes substantial financial, technical, and operational risk in the project (Figure 4.6). The government agency contracts with a private sector consortium to develop, build, maintain, and operate an infrastructure project for a contracted time. The addition of more players can complicate the management and lead to additional problems among the players but, if managed correctly, the PPP provides Owners with a powerful tool to deliver needed public assets without having to find the funds for these projects. In general, the dos and don'ts of managing Design-Build projects contained in this book can be directly applied to managing PPPs.

There are several advantages and disadvantages of each contracting method chosen for a given project. Table 4.1 highlights several advantages and disadvantages for each.

INTERACTION WITH OWNER AND OWNER'S ENGINEERING CONSULTANT

It is imperative to have a good working relationship among all the parties of the Design-Build team. Clear communication is essential. The parties' roles and relationships should be clearly spelled out in the contract(s) between the Owner and the Design-Build entity and in the teaming agreement among the parties of the Design-Build team. These roles are not the same as Design-Bid-Build, and unless this is understood, disputes can arise when expectations are not met.

The Owner must be involved in the communication flow of the project; otherwise, the process breaks down or impedes the progress of the work. Traditionally, under Design-Bid-Build, the Owner has easy access to the Engineer, which acts as the representative for the Owner and has an ethical responsibility to act in accordance with statutes and regulations. In Design-Build, the Designer has responsibilities mandated by statutes and regulations but has a contractual responsibility to the team's success and finances. For this reason, the Owner may not have the same direct access to the Designer that it would in a Design-Bid-Build project, where it employs the Engineer directly.

Recognizing these restrictions, communication between the Design-Build team, the Owner, and the Owner's Engineering Consultant needs to be addressed in the contract between the Owner and Design-Build team. Clear and timely communication is essential to a successful project.

TABLE 4-1 Advantages and disadvantages of contracting methods

	Advantages	Disadvantages
Overall Design-Build	• Total project cost is known earlier. • Flexibility is greater in selecting team. • Single point of responsibility exists. • Project delivery is faster. • Teaming is fostered. • Owner's contract administration is simplified. • More risks can be assigned to the Design-Builder. • Claims are fewer against Owner for faulty design. • Insurer and surety have potential advantages, and Design-Builder has more control of risk. • Owner-controlled insurance policies provide potentially lower insurance costs. • Greater use of partnering exists.	• Different and unfamiliar system of checks and balances is required. • Project scope must be well defined. • Owner surrenders value engineering participation. • Owner surrenders detailed project control. • Owner/Contractor/Designer roles are different, thus an unfamiliar way of doing business. • Owner has less say in materials selection and means and methods of construction. • Public perception is that Design-Builder is less sensitive to the public's interest than Owner-controlled Design-Bid-Build. • Surety risks are higher since Design-Builder has greater responsibility for project performance—both design and construction.
Contractor-Led	• Financial strength (net worth and bonding capacity) is greater. • Contractor is already structured to assume risk. • Contractor is experienced in estimating and bidding hard-dollar construction. • Contractor is experienced in dealing with unions and craft labor. • Contractor has risk-taking mindset. • Design has benefit of more constructability review.	• Owner has less control of design. • Owner required to clearly specify design performance upfront (in proposal stage). • Less developed design at the bid stage may increase uncertainties and risks even to Owner. • Contractor has greater interest in initial cost and not necessarily long-term performance.
Designer-Led	• Designer has more sophisticated technical understanding of design challenges. • Designer may offer expertise and experience in subsurface structures not normally found in contractor organizations. • Owner may feel more in control of quality with a Designer-led team. • Public may be assured of more attention to design concerns and public safety.	• Design firms may need to form new corporate entity to assume risks, obtain appropriate insurance, and licenses. • Design firms are not experienced in estimating, bidding, subcontracting, or managing construction work. • Designer may not have ability to bond construction work. • Owner may have to assume a more active role in resolving construction issues. • Insurance and surety costs may increase because most Design firms do not have a track record in overall project delivery.
Designer/ Contractor joint venture	• If properly structured, joint venture can harness the strengths of both Designers and Contractors in undertaking technically challenging subsurface work. • Joint venture takes advantage of true partnership between Designer and Contractor. • Technical expertise is more likely to remain involved throughout construction. • It is more likely that sound technical solutions will be identified and implemented.	• Both parties are subject to joint and several liabilities. • If a new joint venture, the entity will not have a track record. • More complex contracts are normally required to limit liability and control construction risks. • Joint ventures are more difficult to manage. • More sophisticated Owner organization is required to interact with Designer/ Contractor joint venture.

TEAM COORDINATION RESPONSIBILITIES

One of the significant advantages of Design-Build is the generally less adversarial relationships between the project's participants. The parties tend to view themselves as partners, even though contractually the onus of performance is still squarely on the Design-Builder. Quite often the Owner will co-locate its project management staff in the same office as the Design-Builder in order to facilitate communication and day-to-day coordination. It is incumbent upon all Design-Build parties and other major stakeholders to ensure complete communication and coordination of all phases of the project from inception through completion. Use of formal partnering work-shops in the early stages and regular partnering reviews throughout the project helps facilitate this coordination and resolution of issues at the earliest possible time and at the lowest level in the organization as possible.

CONCLUSION AND RECOMMENDATIONS

Design-Build is becoming a more common and effective delivery method. Although it has been employed on a few subsurface projects, more Owners are becoming interested in trying out this approach. Evolving experience indicates that Design-Build is becoming useful and desirable as a project delivery vehicle. The reasons Owners cite for their interest in Design-Build include faster project delivery time, a single point of responsibility, and potentially lower overall project costs (mostly as a result of the faster project delivery).

Several factors inhibit more widespread use of Design-Build on subsurface projects:

- Prevailing state and local laws mandate the more traditional Design-Bid-Build approach. These laws came about to prevent abuses of Contractor selection or award and to separate the Engineer's role from both the Contractor's role and the Engineer's duty to the public good.

- Most subsurface projects are public rather than private undertakings. Accordingly, a great number of stakeholders must be considered, consulted, satisfied, and updated during the planning, design, and construction of public subsurface projects. Owners are reluctant and not accustomed to giving up control of these contacts to others.

- The geotechnical risk is higher at the time of bid. If the Contractor does not have adequate subsurface information (boring logs, Geotechnical Data Reports, and Geotechnical Base-line Reports) at the time the price must be submitted (and committed to), the proposed price must include a higher level of contingency to cover unknown soil conditions.

- Because there are few examples where Design-Build has been used on subsurface projects, Owners are hesitant to commit funds on an approach that lacks a favorable track record.

The roles, responsibilities, liabilities, and demands that each participant assumes in a Design-Build project vary from the traditional Design-Bid-Build approach. Each participant must understand these differences and be willing to accept the new roles, so that the process can proceed smoothly without the players returning to more familiar but potentially incorrect behavior.

Many Design-Build teaming structures can be used with this delivery method:

- Contractor-led is the most popular arrangement so far because Contractors are more familiar with dealing with risk, are usually more financially able to provide the bonding for the project, and are more familiar with managing labor and the actual construction process.

- Designer-led is gaining some popularity because it can offer Owners some comfort with the more familiar role of dealing with the Designer as the main point of contact with the Contractor in a subordinate role.

- The Contractor/Designer joint venture is less favorable due to the joint liability imposed on both parties. The Designer is not usually structured for, nor comfortable with, actual construction risks.

The authors believe that the Contractor-led model is best suited for subsurface projects. The primary reasons for this belief is that Contractors are more comfortable with risk, have better experience in actual subsurface construction, and are usually financially better able to deal with the uncertainties or unknowns that always seem to reveal themselves during subsurface projects. Typically, these projects are labor-intensive, utilize expensive equipment, and are susceptible to large cost overruns when the operations are interrupted. Contractors are more familiar with working out these disruptions, especially when they are teamed with a Designer in a cooperative, mutually supportive effort, instead of the more traditional adversarial roles encountered in Design-Bid-Build.

Regardless of the structure utilized in the Design-Build team, all parties must pledge to communicate openly and fully to achieve a successful outcome. They must work toward the successful completion of the project with an understanding that each is entitled to its measure of success, profits for the Design-Builder, and a project that performs the desired function delivered on time and at a cost as low as possible for the Owner. One key to success in implementing Design-Build is the level of trust among team members. This can be accomplished by employing and embracing a partnering atmosphere. (See Chapter 10 for a further discussion of partnering among Design-Build team members).

Recommendation 4-1: *The Owner must clearly define the project and develop a detailed scope of services that it desires. Not having a clearly defined scope can lead to misunderstandings, disputes, and mistrust among the parties. In addition, an unclear scope will inevitably lead to scope creep, additional costs, and a disappointed Owner and Design-Build team.*

Recommendation 4-2: *There must be communication between the Design-Build Team and the Owner during all phases of the project. The expectations of each party must be clearly communicated to the other.*

Recommendation 4-3: *Formal partnering should be adopted by the Owner and the Design-Build team as a way to handle the inevitable conflicts that will develop during the life of the Design-Build project.*

Recommendation 4-4: *The Design-Build team must have a teaming arrangement that clearly spells out the roles and responsibilities of each party, the deliverables and schedule for these, the relative financial participation and risks or rewards of each party, and how internal disputes will be resolved.*

Recommendation 4-5: *Each party to the Design-Build procedure needs to have a voice at the negotiation table so there is clear understanding of what is being asked and agreed upon and are no surprises.*

Recommendation 4-6: *A joint Contractor/Designer risk assessment is an effective way to get both parties on the same side to avoid the "us versus them" trap.*

Recommendation 4-7: *A review by an outside insurance expert of the insurance products, project requirements, risk matrix, and teaming agreement is advised.*

Recommendation 4-8: *The Design-Build contract provisions should clearly delineate each party's risk responsibilities and include risk-sharing clauses (e.g., risk-sharing pools, provisional sums, and/or Contractor incentives for managing risk effectively). The party best able to manage or mitigate risks should be contractually responsible for them. In areas where risks are unknown or poorly defined at the time of bidding (such as unknown or poorly defined geotechnical conditions), provisions for price adjustment should be built into fixed-price–type contracts to help share cost-risk and provide equitable adjustment once such conditions or risks are better defined. This will minimize the need for unnecessarily large cost contingency in the bid price.*

REFERENCES

Friedlander, M. 1997. Entering a Design-Build agreement? A plan of action. *Consulting-Specifying Engineer*, August, p. 25.

———. 2000. Designer-led Design-Build: why it works for contractors. *Construction Lawyer*, January, p. 29.

Smith, G. 1996. Design-Build concept expands. *Garry Smith's Office*. www.gsoexpert.com/designbuild.htm. Accessed November 2009.

CHAPTER 5

Procuring the Design-Build Team

Charles Button, P.E.

Thom L. Neff, P.E., PhD
President, OckhamKonsult, Boston, Mass.

INTRODUCTION

The success of any major subsurface construction project depends upon many factors. For a Design-Build subsurface project, these factors are not new, but their inter-relationships demand significant changes in how the Owner procures the project team. A brief list of important factors for success includes

- Nature and complexity of the project;
- Sophistication and experience level of all involved parties;
- Methods and approach utilized;
- Schedule, funding, and third-party constraints;
- Level of project uncertainty;
- Ability of all parties to communicate and collaborate effectively;
- Regulatory constraints; and
- How completely and fairly the project risks (uncertainties) are defined, allocated, managed, and controlled during the entire process from procurement through construction and operation.

Experience with the Design-Build project delivery system, even for subsurface projects, has increased in recent years, so that we have some precedents to guide future applications. Some Owners prefer this method for its single-point responsibility, as well as for additional positive factors discussed in detail in other chapters of this book. Even though Owners' experience with subsurface projects and Design-Build may be limited, they are familiar with their own specific rules and regulations, and typical project constraints. It thus becomes critical for them, once they have determined they want to employ the Design-Build process on a particular project, to get it right. It's the Owner's money, and the Owner knows what it wants in terms of the completed facility's performance. The Owner must have adequate control over the entire process, starting with procurement of the Design-Build team.

This chapter will describe the procurement process, in general, and the most important issues required for a successful Design-Build procurement, along with their critical points of contention. It will provide a concise summary of recommended best practices, based upon the authors' personal experience.

PROCUREMENT PROCESS—GENERAL

An in-depth understanding of applicable legal requirements, as well as a familiarity with formal procurement methodologies and the policies on which they are based, are essential in undertaking any large or complex contract procurement, such as a Design-Build subsurface project. There are many published materials on all aspects of formal contract procurement; helpful treatises on source selection include *Formation of Government Contracts* (Nash and Cibinic 1998) and *Competitive Negotiation: The Source Selection Process* (Nash et al. 1999). The basic principles elucidated in these books do not change in the Design-Build context. Published materials dealing with Design-Build construction contracting in the public sector include *Design-Build Contracting Handbook* (Cushman and Loulakis 2001). In this chapter, we assume the reader's familiarity with basic procurement methods and techniques commonly used in government contracting.

As in Design-Bid-Build and other project delivery approaches, a number of procurement methods can be used for Design-Build contracting. Most large subsurface construction works are undertaken by public agencies as part of major infrastructure projects. Therefore, most contract awards for these projects will be subject to formal procurement procedures imposed by law or regulation, which may or may not deal explicitly with Design-Build. In many jurisdictions, use of Design-Build by state or municipal agencies has been permitted only through special legislative authority to avoid application of existing laws that reflect the traditional Design-Bid-Build approach, which usually requires separate procurement procedures for engagement of Engineers (typically qualifications-based) and Contractors (typically low-bid based on 100% design). Even where applicable procurement law allows Design-Build contracting, laws concerning professional licensing may constrain or even preclude Design-Build. Projects that receive federal funding may be subject to additional or conflicting procurement requirements. Where existing law does not currently permit Design-Build, there may be the potential for changes in laws or regulations to permit Design-Build contracting or to create greater flexibility in procurement methods.

Regardless of the applicable legal procurement framework, advance planning, well-thought-out selection procedures, and carefully prepared documentation with respect to each stage of the process are critical. The risk of an award protest on a major Design-Build subsurface project is always present, because

- The investment of time and money necessary to develop and submit a proposal is substantial;
- The contracts are large and potentially lucrative; and
- Design-Build contracts are awarded using nontraditional procurement approaches such as best-value selection, which may be perceived as subjective and, therefore, potentially unfair.

If a protest is brought, and, more importantly, in order to make the right selection decision, an agency must invest the time and effort to structure and implement an effective and defensible procurement process. Agencies accustomed to procuring construction work on a low-bid basis find that the procurement process for a major Design-Build project requires a significantly greater effort in planning, preparation, and execution, particularly with the agency's first experience with this contracting method.

In a Design-Build subsurface project, each proposer generally must prepare and submit extensive information and analysis explaining its proposed design approach and construction means and methods, as well as price and other proposal materials, which the Owner must evaluate and compare through a process that will typically consume six or more months. In large complex projects, the total time period required for the procurement process will often exceed one year. Owners that are interested in Design-Build primarily to achieve a shorter project schedule should factor in the longer time required for the procurement process itself.

Owner Choices and Process Control

As noted, the legal requirements applicable to a Design-Build procurement may vary substantially in different jurisdictions. In addition, every major Design-Build project faces a unique set of circumstances that may impact the procurement approach. In general, Design-Build does not lend itself to a simple low-bid selection approach. The Owner in a Design-Build project often relinquishes some degree of control over the design and construction process to the Design-Builder, resulting in a loss of checks and balances under the Design-Bid-Build method that can help protect the Owner in the design and construction of the project consistent with the Owner's programmatic and contract document requirements. The Owner's greater vulnerability requires a correspondingly higher level of trust and confidence in the experience and reliability of the Design-Build team. These concerns are particularly important in the case of Design-Build subsurface projects, where the potential risks arising from unforeseen conditions or problems in the design or construction methodology are greater and have more serious consequences, and construction deficiencies may be difficult to detect or remedy after completion of the subsurface work.

These considerations often outweigh even significant differences in proposal prices, which experienced Owners recognize as only the starting point for the final project cost. For these reasons, a *best-value* selection process, in which selection is based on both price and nonprice criteria, can be preferable for Design-Build subsurface projects. (In this chapter, the term *best-value* is used to indicate the procurement strategy now referred to in the Federal Acquisition Regulations [FAR] as the "trade-off process." Best-value selection and the legal underpinnings of this approach are discussed in Nash et al. 1999.)

Various procedures can be employed to implement best-value selection. It is worth noting that private entities use Design-Build much more commonly than do public agencies, and that a number of new information technology and risk management techniques are now available to better manage and control complex Design-Build projects. Some advocates of Design-Build consider that the Owner's perceived loss of control in the Design-Build process can be largely mitigated through better strategic planning in the procurement and preliminary design phases.

Piecemeal experimentation with Design-Build for construction in the federal arena has been replaced in recent years by a uniform procurement approach directed by the Clinger-Cohen Act of 1996 and implementing regulations (see FAR, Subpart 36.3 and Part 15). Federal agencies opting to use Design-Build must follow a two-phase process for contract procurement. During the first phase, interested parties submit proposals addressing their experience, qualifications, and general technical approach to the project; price or cost information may not be submitted in this phase. Based on an evaluation of the phase-one proposal, including an investigation of the proposer's performance on past projects, the agency selects a short list of proposers (generally not more than five) that are invited to submit a phase-two proposal, which includes more detailed descriptions of the proposer's design concepts and technical approach and proposed price.

The contract award is generally based on the best-value selection approach. Alternatively, the *lowest-price–technically-acceptable* approach can be used, pursuant to which the contract is awarded to the lowest-price proposer found acceptable with respect to established technical and other nonprice criteria, which are applied on a pass/fail basis. The solicitation documents must clearly notify proposers if the lowest-price–technically-acceptable approach will be used. The federal two-phase procurement method is discussed in more detail in various sources, including Didier 1998 and Cushman and Loulakis 2001, Chapter 9.

The primary purpose of pre-qualification, short-listing, or the two-phase process is to reduce the competitive field to a manageable number of well-qualified proposers. This benefits the Owner by reducing the effort required to evaluate proposals, which is a substantial task in the case of a major project, and it benefits proposers by limiting the number of teams that will incur the time and expense of preparing a full technical and price proposal. These are significant concerns in many Design-Build projects, and experts often recommend use of pre-qualification in Design-Build procurements. In the case of a major subsurface Design-Build project, however, the need to limit competition is generally not a concern, because the universe of players able to compete for and perform these contracts is limited. Nevertheless, some Owners feel that pre-qualification is a valuable gauge of proposer interest in the project, and that the proposers prefer pre-qualification.

Submission by a team to a request for quotation (RFQ) does not ensure, however, that the team will respond to the subsequent request for proposal (RFP), and industry interest in the project can be gauged by issuance of a request for expression of interest and other less formal methods. From the proposer's standpoint, a pre-qualification phase will eliminate unqualified potential low-ball bidders, which may be an advantage in many circumstances. However, in the case of a major subsurface Design-Build project, the time and expense of participating in the procurement is itself somewhat self-selecting. An unqualified proposer is unlikely to expend the time and money necessary to prepare and submit a serious proposal.

The potential disadvantage of pre-qualification is that a firm or team that does not respond to the RFQ, for whatever reason, will be ineligible to respond to the RFP, no matter how strong its qualifications. Legal issues can also arise if an otherwise qualified party responding to the RFQ does not comply with the formal requirements, or if questions are raised as to which member(s) of the proposer team must be pre-qualified, in the case where teams are not finalized when qualifications statements are due or if changes in principal team members subsequently occur.

Best-Value Versus Lowest-Price Selection Options

Whatever procurement approach is used, a best-value selection method is usually preferable to the lowest-price–technically-acceptable method. Even if two proposer teams are indistinguishable with respect to qualifications and prior experience, and have each submitted technical proposals that reflect acceptable approaches to the design and construction of the project, the evaluation team may consider differences in their approaches to be as or more important than differences in price. In tunneling and other subsurface work, there are often several alternative approaches to design and construction. One of the potential benefits of Design-Build in this context is the opportunity to take advantage of the innovation and technical evolution that characterizes the subsurface industry. A procuring agency may receive competing proposals that are very different but technically acceptable—that is, capable of successfully constructing the project. The lowest-price–technically-acceptable method may simplify the evaluation process in some cases by eliminating the need to perform a price/technical trade-off to determine the best value. However, it may also leave the procuring agency forced to accept a proposal that offers only modest cost savings and is based on a technical approach that would not be the agency's first choice with respect to other important factors.

For example, assuming that the contract places the risk of unknown or differing subsurface conditions on the Owner, the Owner may properly be concerned about the extent to which different approaches to design and construction may mitigate or exacerbate that risk. The potential economic exposure to the Owner may be substantial. The fact that one proposer's approach may increase the likelihood of claims for differing site conditions, however, does not mean the approach is not technically acceptable. If the proposer is not eliminated as technically unacceptable and submits the lowest price, it will be selected for award under the lowest-price–technically-acceptable method, even if the evaluation team believes that its approach is more likely to result in claims for increased cost and time extension, ultimately resulting in higher costs to the Owner.

Under the best-value method, if the proposer that is rated highest as to nonprice evaluation factors does not have the lowest price, the Owner is required to judge whether the advantages offered by the highest-rated proposer outweighs the price premium that must be paid to obtain those advantages. This trade-off analysis must be carefully documented, and, depending on the circumstances, may be difficult and add time to the evaluation schedule. An award to other than the lowest-price proposer may increase the chances of a protest or challenge to the award. However, under the lowest-price approach, protests may also arise as to whether a proposal selected for the award or a proposal that has been eliminated is in fact technically acceptable under the established criteria.

As discussed in more detail in other chapters, the continuing debate concerning Design-Build centers on the extent to which the Owner sacrifices control under Design-Build. To make Design-Build work, Owners must look for ways to maintain control over critical elements of the project with minimum sacrifice of Design-Build's advantages. Selecting the proposer that best satisfies all the Owner's criteria—and not just price—may not be an easy task in every case, but the ability to make that judgment allows the Owner to retain a measure of control over the final design approach and construction means and methods with minimal impact on the procurement schedule and no impact on proposer creativity or any of the advantages of Design-Build.

The overall procurement approach that will be the most feasible and effective for any given project will depend upon a number of factors, and many of the practical issues that arise in conducting the procurement will be the same under any of the approaches discussed. If legal requirements do not permit best-value selection and require award to the lowest bidder but permit pre-qualification, then a variant of the two-phase approach would be preferable to a simple low-bid procurement. By limiting the pre-qualified short list to firms that are acceptable with respect to prior experience, qualifications, and basic technical approach, the risks to the Owner arising out of a lowest-price award are somewhat mitigated. Under such an approach, it becomes important to focus adequate attention on the pre-qualification phase by establishing minimum requirements for qualifications and experience that will ensure qualified proposers without unduly restricting competition. In state and local procurements, legal requirements may limit or preclude use of the lowest-price–technically-acceptable approach as permitted under federal law. In such cases, once a Design-Build team has been found to be pre-qualified, it may not be possible to eliminate them at the proposal stage as technically unacceptable. Therefore, it may be advisable under this approach to require a more developed statement of the proposer's technical approach at the pre-qualification stage, if that is permitted.

On the other hand, if the best-value selection is permitted, a pre-qualification or a shortlisting phase may not be necessary (unless it is legally required) in the use of Design-Build subsurface procurements, for the reasons already discussed. Whether or not pre-qualification is used, there are clear advantages to employing a multiple-stage process that includes technical and price proposals, discussions or negotiations with proposers, revision of RFP requirements if necessary, and revised proposals, as discussed in the next section.

Implementing the procurement process for a Design-Build subsurface project involves consideration of many of the same issues that arise in connection with any major construction procurement. Differences exist, however, in recognizing and responding to these issues in a subsurface Design-Build project. We assume the reader's familiarity with the basics of procurement and will not review or comment on all the steps that are followed in a typical procurement or the many variations. Rather, we selected what we believe to be the more important issues in subsurface Design-Build, which follow.

PRELIMINARY DESIGN

One of the earliest decisions the Owner must make is the level of preliminary design that will be completed for the Design-Build bid package. For all but the most sophisticated entities, the typical Owner will have to enhance its in-house staff with a number of consultants (individuals and firms) to properly prepare for a Design-Build procurement. It is prudent for these consultants to have had some experience with the Design-Build process on similar projects. In general, these consultants will retain a significant role throughout the entire project, advising the Owner through final design and construction, even in some cases serving as the Owner's Construction Manager in a reduced but critical role for reviews and approvals of the Design-Build team's work progress. The Owner knows better than any other project player what is required in a finished facility: how the facility will operate and perform for its design life. The Owner will work closely with its consultants to describe carefully and completely the desired facility, while leaving maximum flexibility to the selected Design-Build team in how the facility will be created. In this manner, the primary potential effectiveness of the Design-Build approach can be achieved.

In general, to provide the bidders enough information to create their proposal in a reasonable amount of time, the preliminary design will have to be taken to at least the 30% level. This will permit the Owner to define any elements that are crucial to operation and maintenance of the completed facility, and to give reasonable guidance to other important design and construction elements: for example, third-party issues, utilities, environmental constraints, and integration with the Owner's existing adjacent or related facilities. The preliminary design will include drawings, design criteria, and certain critical specifications or unique equipment that the Owner must have for its specific operational needs. Although it is assumed that some prescriptive elements will be included, it is recognized that the more prescriptive the preliminary design, the less creative the Design-Build team can be in cost-effectively carrying out the work. In some cases, an Owner may go so far as to dictate the approach to be used in the final design; however, this will effectively reduce the Design-Build team's ability to optimize its inherent strengths.

The Owner should not lose sight of the fact that it can retain sufficient control to ensure that the end result—the completed project—meets its specifications and needs. To the extent the Owner leaves the means and methods to the Design-Build team, it will more likely get a project that is both cost-effective and timely. The geotechnical aspects of the preliminary design represent an exception to the general approach for Design-Build, because this issue for subsurface projects will likely represent the single largest source of uncertainty for the project. Because of the specialized nature of geotechnical engineering, the Owner will likely hire a specialty firm for this work and use a local firm because of the site-specific nature of geotechnical issues. If this firm has Design-Build experience, it will be a bonus, but, if not, an individual geotechnical consultant can fill this gap.

It would be prudent to take the geotechnical design far beyond the 20%–30% level of the general preliminary design effort. Some would even lobby for a complete geotechnical design that includes sufficient borings, samples, and testing to produce a Geotechnical Data Report (GDR) and a Geotechnical Baseline Report (GBR) for use by the Design-Build team. A full discussion of the GDR and GBR, as well as the differing site conditions clause, appears in Chapter 9. The intent of these documents, as with the preliminary design in general, is to provide potential bidders with a clear picture of what they will have to deal with as they complete the final design and construction. An added advantage of doing most, if not all, of the geotechnical work upfront is to dramatically reduce the actual procurement period. The Owner has the luxury of strategically planning the entire sequence to meet its overall schedule and can do this both to save time and to present a bid package containing the lowest amount of uncertainty (and resulting risk). It remains a given that the Design-Build team will search out and price uncertainty and risk accordingly. There will be no free lunch. Risk definition and allocation are discussed in more detail in Chapter 3, but it is clear that a comprehensive effort by the Owner and its consultants in the preliminary design effort will go a long way toward receiving fair and responsive bids.

A further benefit and efficiency to the process can result from a well-planned and executed geotechnical program that includes environmental sampling and testing. Often, contaminated groundwater, soil, and rock, as well as the presence of subsurface manmade obstructions, can be detected and defined during the same boring and sampling program for a modest additional cost. The testing and analysis of the resulting environmental data are, of course, an added expense, but considered a good investment because it will further indicate to the bidders that the Owner has made reasonable efforts to define uncertainties in the bid package. The Owner must, however, give the winning bidder an option to obtain additional subsurface information if it so desires, and any resulting information must be incorporated into the appropriate GDR, GBR, and other contract documents prior to executing the work.

OWNER'S PROCUREMENT TEAM

Because of the complexities and risks presented in large subsurface projects, an Owner must include on its procurement team representatives with expertise in subsurface construction who can evaluate the alternative design and construction approaches proposed by the Design-Build teams. The Owner typically will have engaged geotechnical and design engineers early in the process to perform subsurface investigations and develop a preliminary design. Appropriate members of that design team should be part of the Owner's evaluation and selection team. In most cases, it is advisable for the evaluation team to also include other experts in geotechnical matters who did not participate in preparing the preliminary design. New subsurface construction techniques have continued to evolve in recent years, and evaluation team members must be knowledgeable about the current state of the art. At the same time, breadth of experience in dealing with practical geotechnical engineering issues, such as ground behavior, remains essential in evaluating the advantages, disadvantages, and risks for different design and construction approaches.

Effort expended by an Owner to identify and recruit qualified consultants to participate in the selection process is a sound investment. In the event of a legal challenge, reviewing courts or boards will generally give substantial deference to the Owner's reliance on the informed judgments and recommendations of independent experts with experience in similar situations. More importantly, the selection decision on a major subsurface Design-Build contract award is a complicated task with high stakes riding on the outcome. It is critical that the agency's decision-makers have available the most objective, comprehensive, and informed analysis possible. If the Owner's evaluation team is to include engineering professionals who did not participate in the initial geotechnical analysis and preparation of the preliminary design, those individuals should be brought on board as early as possible, with the commitment to spend the time necessary to familiarize themselves with the preliminary design and available geotechnical information prior to receipt of proposals.

In general, the Owner will not want the engineers who prepared the preliminary design to be eligible to compete for the Design-Build contract or participate in any way on a proposer team because of the potential for conflict of interest and the perception of unfair advantage vis-à-vis other proposers. Whether or not such ineligibility is mandated by applicable law, it should be expressly provided for, both in the contract between the Owner and the engineer for the preliminary design work and in the solicitation documents for the Design-Build contract.

In any major procurement, the engineers on the Owner's team must work closely with legal counsel and other parties involved in preparing the solicitation documents to ensure that the documents include all information determined to be relevant, explain clearly what materials and information must be submitted by proposers, and state all evaluation criteria that the evaluators will apply. Detailed instructions and guidelines for conduct of the evaluation process should be prepared and distributed to all members of the evaluation team. These instructions can alert the evaluation team to applicable legal requirements, confidentiality and conflict-of-interest concerns, rating systems, documentation requirements, and other matters. The evaluation criteria should be understood by all members of the Owner's evaluation team. It is also important for key team members to review and understand material terms and conditions of the contract, including the allocation of risks for differing site conditions, liability for preliminary and final designs, and means and methods of construction. An understanding of the contract requirements will aid in the evaluation team's assessment of the strengths or weaknesses of the proposal.

PROPOSAL REQUIREMENTS

Like most of the main topics in this chapter, the proposal requirements are closely interrelated. The Owner and its consultants need to realize that the entire strategic planning of the procurement process should focus on clear, concise elements that do not put an undue burden on the proposers, but at the same time permits an unbiased and efficient review and selection process. Some of the elements that will determine what the RFP and the resulting proposals actually contain are

- Applicable local, state, and federal procurement regulations and other legal issues;
- Significance of the bid price in the selection;
- Whether best value is used in the selection process;
- Level of detail in a pre-qualification process;
- Whether a single-stage or a multiple-stage proposal process is used;
- Whether revisions or direct communication with the proposers are permitted after the selection of the approved bidding teams;
- Whether some proposal terms of the proposers will be incorporated into the contract documents, especially as they relate to constructability or construction sequencing issues;
- How to handle proprietary materials, equipment, and/or processes of potential bidders;

- Whether additional geotechnical data obtained by the winning proposer will be incorporated into the contract documents (revised GDR or GBR); and
- Whether technical and price proposals are submitted separately.

The RFP must avoid the following two extremes: (1) a winning low-ball bid by an unqualified team, and (2) a long, drawn-out negotiation that results in a bid protest by one or more bidders. Thus, a clear, concise RFP that lays out exactly what the Owner wants to see in the proposals and what is and is not permitted will ensure a smooth process. The Owner and its consultants are in charge and will ultimately determine how the procurement flows. In a general sense, it is always desirable to have Design-Build team players that have some previous Design-Build experience and, further, that have previously worked together.

The RFP should seek details on

- Actual functioning of the Design-Build team (one or more locations),
- The team's approach to the final design and construction,
- Integration and coordination of the design team with the construction staff,
- Total project schedule with all critical milestones,
- Breakdown of all pricing,
- Management and control of uncertainty and risk,
- Execution and control of specific means and methods, and
- Implementation of submittal and review and approval process.

Real-world examples of RFPs have included most of these elements, but it is the authors' opinion that a fair and reasonable approach by the Owner and its consultants, as outlined in the preceding list, will help to clearly define uncertainty and risk and to create contract documents that allocate those risks to the various project participants in a fair and reasonable manner.

Because of the complex nature of subsurface design and construction, it is possible that a given procurement could wind up with a single qualified bidder for the work, and the Owner should be prepared for such an outcome. In cases where this happened, the Owner went ahead with the work because the bid was judged to be good value. Communication is the key element both before and during the entire proposal process. The more clearly the potential bidders understand exactly what the Owner wants, the better chance that it will happen. Prior to issuing an RFP, it would be prudent to have face-to-face meetings with all potential bidders to openly discuss the details of the entire procurement process. In addition, the Owner and its consultants are likely to learn something significant during the RFP and proposal submittal and evaluation process, which will benefit the project by incorporation into the contract documents.

EVALUATION CRITERIA AND SELECTION DECISION

As in any major procurement, the Owner's team must develop and carefully articulate the evaluation criteria that will be used to select the Design-Build team. One of the challenges from using the best-value approach rather than the low-bid method is the greater effort required to develop and apply the evaluation criteria and rating system. Use of the best-value approach does not mean that all evaluation criteria must be comparative; criteria applied on a pass/fail basis may be combined with comparative criteria.

Some criteria will focus on the proposer's qualifications and experience. One of the problems common to all Design-Build procurements is that instead of selecting the best (most qualified, capable, reliable, and experienced) Engineer and Contractor for the job, the Owner must select the Design-Build team, including both design and construction firms. Assuming that the Owner can identify the best Engineer and the best Contractor among those competing for a contract, those two firms may not be on the same team. In addition, the most highly qualified team may not propose the best design approach, and the highest-ranking team, with respect to technical and experience criteria, may not propose the best price. Assuming that the best-value method will be used, the solicitation documents should reflect that the Owner will make judgments and trade-offs in reaching a final selection decision based on a combination of price and nonprice factors.

The evaluation of information concerning a proposer's past performance has been an increasingly important—and controversial—issue in government contract procurement in recent years. Because of the Owner's greater reliance on the responsibility and integrity of the Design-Builder under the Design-Build approach, it is suggested that significant weight be placed on past performance in the case of a subsurface Design-Build procurement. Legal and practical issues concerning evaluation of past performance are discussed in Cushman and Loulakis 2001, Chapter 6. As the industry gains more experience with Design-Build, many Owners may attach greater importance to a proposer team's experience or lack of experience working together previously on a Design-Build basis.

In Design-Build contracting there is a natural and potentially constructive tension between the desire of the Engineer to develop the highest-quality, safest design and the Contractor's desire to develop the most cost-effective design. In subsurface construction, the stakes in this game are high because the ability to successfully complete construction on schedule, or even ensure the safety of the construction workers and the public, may depend upon the judgments made during the course of final design development. Contracting entities, whether Engineers or Contractors, which have not participated in Design-Build projects previously are unlikely to fully appreciate the nature of this tension and how it will manifest itself. Even when prior experience with Design-Build contracting exists, if team members have not worked together before, there is a risk they will encounter difficulties in managing this tension, with the potential for delays or quality problems.

Teamwork is an essential attribute of successful Design-Build construction. Although proponents of Design-Build often hearken back to the master builder of times long past, for various cultural and economic reasons, that model has no counterpart in modern industry practice. However, joint ventures of Designers and Contractors that understand and respect each other's contributions can and do exist, and have completed large and complex construction projects around the world. It is important for the Owner to determine, to the extent possible based on evidence of past experience, references, and discussions with proposers, whether a proposer team will function well. The RFP should require proposers to provide detailed descriptions of how the Design-Build team will be managed and what specific mechanisms will be implemented to foster an effective working relationship between the Designer and Contractor.

Because of the typically large disparity between the Designer and the Contractor with respect to financial participation in the venture, and in some cases because of cultural mismatch, one of the most common destructive pressures on the Design-Build team is the tendency of the Contractor, motivated by the categorical imperative of controlling cost and schedule, to run roughshod over the Designer and its schedule, budget, and design objectives. Although assurances from a proposer team that this will not occur are of little value, the Owner should insist upon specific assurances in its proposal (which should then be incorporated specifically into the contract) that the proposer team recognizes these issues and has a coherent and credible plan to address them, including adequate scope definition, level of effort and quality control procedures for the design work, sufficient allocation of funds in the schedule of values for the design work, and allowance of sufficient time in the schedule for iterative development of the final design.

As previously indicated, successful experience of the team members on previous Design-Build projects is often the most convincing evidence. Although the weight given to the various evaluation criteria will depend on many factors, significant weight should be given to this factor, which, in the authors' experience, is one of the more important predictors of a Design-Build project's success. The Owner's first and best opportunity (and in many cases its only opportunity) to promote this good working relationship is in the selection decision.

In addition to the effectiveness of the Design-Build team relationship, other frequent causes of problems on subsurface projects are the final design approach and construction means and methods chosen by the Design-Builder, and claims for differing site conditions. The evaluation criteria should include factors by which the Owner can differentiate among proposers based on the relative risks posed by their approaches to these issues, as determined by the evaluation team.

Owners must evaluate proposals based on the evaluation criteria stated in the RFP. Although notational scoring or rating systems may or may not be required by applicable procurement law, in a complex procurement such as a Design-Build subsurface project, a notational scoring system, usually based on adjective or point scoring, is almost always used. The Owner and its consultants should decide on the scoring system, including the relative weights of the evaluation criteria, in advance of receiving proposals. A determination must be made as to how much information about the evaluation process and the scoring system should be disclosed to proposers in the RFP. Many different proposal rating/ranking systems can be used; a discussion of the alternatives will not be included here. Evaluation criteria, scoring methods, and the process of evaluation and ranking of proposals under the best-value approach are discussed in many sources, including Nash et al. 1999, Chapters 2, 3, and 6.

CONTRACT PACKAGING

Many of the decisions an Owner needs to make regarding the procurement process are both important and difficult. Whether to advertise the project in a single or multiple contract presents a tough choice. A major reason for this dilemma results from the somewhat unique character of Design-Build projects and the specific case of the even rarer subsurface Design-Build works. Also, subsurface projects are generally perceived to be more risky than conventional works, although better methods to manage risk are now available and are discussed in Chapter 3. In addition, the following factors should be carefully considered before deciding upon the number of contracts that is appropriate for a specific subsurface Design-Build project:

- Local political climate;
- Size and complexity;
- Availability of qualified local design and construction firms to do the work;
- Potential economies of scale (is bigger cheaper?);
- Access and mobilization issues for the winning firms; and
- Perceived risk, in particular, for the subsurface excavation and support portions of the work.

Common arguments for multiple contracts frequently cite more bidders for smaller contracts, more competition, more local participation, and more involvement by the U.S. Department of

Transportation's Disadvantaged Business Enterprise program. The most common argument for a single contract is that there are not that many qualified *and* experienced design and construction firms that have successfully done Design-Build work. One reason it makes sense to have more than one contract is the different nature of the scope of work for the subsurface excavations or openings and the equipment and fitting-out of what goes into the openings. The level of uncertainty is usually also different for these two elements, and thus it might make sense to separate them. The types of contractors that perform these elements are further specialized and look at risk quite differently.

Often not discussed but absolutely critical in this decision is the inescapable fact that, with multiple contracts, the management complexity and resulting costs will rise significantly. With multiple contracts, managing the interfaces between separate contract elements will require great skill and, in itself, may hold potential for claims.

STIPENDS

Major procurements often require proposers to expend significant time and effort preparing proposals and participating in the selection process, and this is certainly true in the case of subsurface Design-Build projects. The effort and expense required to submit a proposal depend in part on the extent of design material required to be included in the proposal.

At a minimum, the proposer is required to determine its approach to final design and construction of the subsurface work and document that approach in sufficient detail to permit evaluation. This requires

- Analysis of the available geotechnical information and preliminary design provided by the Owner;
- Development of alternative solutions for the final design for the work;
- Evaluations of alternatives with the Contractor(s) so as to take advantage of the experience, personnel, and equipment resources of the Contractor(s); and
- Documentation of the selected alternative.

The process requires a substantial commitment of time, mostly by the engineering members of the proposer team. Both Engineers and Contractors are accustomed to incurring time and expense in competing for engagements. However, in traditional projects Engineers do not typically incur the costs of substantive design work of the type required to submit a proposal on a subsurface Design-Build project. Therefore, the upfront investment by the design team in proposing this type of project is far in excess of the marketing expenses incurred in connection with pursuing a traditional design engagement. In a Design-Build context, the Designer's construction partner may compensate the Designer to some extent for the costs of performing the work necessary to submit a proposal. However, such payments typically fall well short of fully compensating the design team for its costs. Although the Owner receives the benefit of the design development performed by the winning Design-Build team, depending on the contract payment schedule, the Design-Builder may not receive significant payments for design costs early in the project.

For these reasons, and in an effort to promote competition, some Owners offer to pay stipends or honoraria to teams that submit proposals in which significant design work is necessary as part of the proposal process. Commentators have reported that stipends generally range from 0.05% to 0.3% of the estimated total project design and construction cost (Robinson et al. 2001). In other cases, Owners have concluded that such payments are unnecessary and do not result in more or better competition on their procurements. As with Contractors, there is a limited universe of engineering firms with the experience and capacity to function as the design engineer on a major subsurface project, and these tend to be established engineering firms. Based on the limited anecdotal evidence, it is not clear whether or how often Contractors or engineering firms with the capacity to perform major subsurface projects are deterred from competing on major projects solely or primarily by the cost of proposal preparation. Even assuming that some firms are in fact dissuaded from participating, paying a stipend to each proposer team sufficient to cover a meaningful portion of its costs could represent a large expense for the Owner, which is difficult to justify in some circumstances. On the other hand, although the Owner may receive no direct benefit from design work performed by unsuccessful proposers, if a stipend actually induces even one additional serious competitor to participate, it has created substantial value for the Owner. Applicable law may preclude payment of stipends in some cases.

If an Owner contemplates the use of a stipend, it should be designed to provide maximum benefit at a minimum cost. For example, a larger stipend awarded only to the second- and third-ranked proposers (or the first- and second-ranked if no contract is awarded), with no stipend to the successful proposers or any other competitors, allows the Owner to control this expense and might help stimulate competition among the most serious industry players. With respect to the successful proposer, Owners should consider providing in the contract payment schedule for a significant early payment to compensate for design work performed during the proposal phase.

One of the concerns with Design-Build contracting from the Owner's perspective is the attenuation of the relationship between Owner and the final design Designer in contrast to the traditional Design-Bid-Build structure. The strong symbol of a significant payment to the design team upon award of the contract, combined with other affirmative steps by the Owner to reinforce the importance of the design team to the outcome of the project, may enhance the potential for a more open and effective relationship between the Owner and the Engineer. In the end, the Contractor will largely call the shots on the Design-Build team, but the perceptions and relationships established in the initial stages of a project can have an impact on the Owner's timely access to key information and its ability to influence design decisions.

The amount of the stipend would reasonably be a function of the extent to which the design is already developed. The number of borings taken and the extent of the data supplied and baselined would affect the Contractor's uncertainty. The detail to which the tunnel boring machine and tunnel lining are specified, ground subsidence is restricted, and the muck characterization and disposal requirements are detailed all should factor in the effort required by the Design-Build team. More information should reasonably reduce the proposer's costs and therefore the stipend.

CONTRACT DOCUMENTS

Having a set of clear, concise but comprehensive contract documents will go a long way toward a successful procurement. The RFP must clearly state to bidders the form of the contract they will enter into, and in some cases this could have a direct bearing on a potential proposer's decision to bid for the work. The contract documents must be consistent with local, state, and applicable federal regulations, especially those that deal specifically with the Design-Build delivery process. Many jurisdictions have yet to address Design-Build, or have done so in an incomplete fashion. Because it is unlikely that existing Design-Bid-Build documents can serve effectively, even with significant revisions, new documents are the prudent choice. Preparing a good set of contract documents is no small undertaking but must be done with care and attention.

The Owner's team that performs this effort should have adequate input from engineering, legal, project management, construction, and even political sources. Local politics can have a significant effect on any major construction project, especially one that employs a relatively new delivery process. Because it is unlikely the Owner will have sufficient internal staff to undertake a Design-Build procurement, it should retain supplemental design and geotechnical consultants that can effectively address all pertinent legal, economic, political, and technical issues. Collectively, this team must identify and resolve any and all items that can lead to misunderstandings in the execution of the contracted work. A partial list of important issues would include

- How the final design will be completed, reviewed, and approved;
- What the Owner's rights are in all reviews and approvals;
- How risk will be fairly and reasonably allocated;
- How quality assurance/quality control (QA/QC) will be achieved;
- What warranties will be expected; and
- What are the limits of liability, where applicable.

Important to keep in mind throughout the process is that, although all the separate elements of a Design-Build contract have been done before by the Owner, the challenge now is the combination of the elements into a single contract. Covering all the interfaces is the real challenge. Some Owners have considered Design-Build an opportunity to transfer all risk to the Design-Builder, but this is viewed by experienced professionals as a fundamental mistake. It's not a fair and reasonable thing to do, and any qualified bidder will let its price reflect its assessment of the risk it is asked to assume. Again, there is no free lunch. Typically, issues covered by the differing site conditions clause belong to the Owner, and those of means and methods to the Contractor. A prudent Owner will go to great lengths in the preliminary design to reduce uncertainties and clearly define remaining risks. If these issues are clearly stated in the contract documents, it will go a long way toward promoting a successful procurement.

PLANNING AND PROMOTING THE PROCUREMENT

Procurement planning must be incorporated into project planning as early as possible. As in any major undertaking, planning starts with the identification of objectives and constraints, which are more numerous and complex in the case of public sector projects than in the private sector. The timing of major project milestones and commitments is likely to be important. The need to fit into a schedule driven by federal funding cycles, state election cycles, or other uncontrollable factors may be among the reasons for considering the Design-Build approach. Federal and state environmental reviews, community impacts, commitments by state or local politicians, construction industry cycles, local construction industry behavior, and a myriad of other factors may impact the decision as to when and how the project procurement will proceed.

One of the important objectives in structuring the procurement is the avoidance of legal or political problems, such as bid protests, or questions as to the integrity of the process or leaking of confidential evaluation information. One of the first steps in planning a major Design-Build procurement is establishment of the legal parameters that will govern the process. This may include seeking special legislation or approvals, which may involve a long lead time. Certain procurement issues, such as contract packaging, may have an impact on the environmental review process or on budgetary planning. For these reasons, procurement considerations should be part of project planning from the concept stage.

A limited number of firms may have the capability to perform major subsurface projects, but not all of these firms may be interested in performing work on a Design-Build basis. Therefore, early and effective industry outreach is important to ensure that as many qualified companies as possible will participate in the procurement. Also, exchanges of information between the procuring agency and industry participants can provide valuable input to the agency about a host of issues, including procurement structure and schedule, contract packaging, approach to preliminary design, contract requirements, and data requirements, at an early stage when the agency can more easily adapt its approach to accommodate industry concerns. Design and construction firms experienced in major subsurface projects are a source of valuable information and experience, and the Owner should take full advantage of this resource.

Legal requirements must be considered when planning outreach activities. The FAR expressly permits numerous types of information exchanges prior to commencement of formal procurement, including not only industry conferences and site visits but also one-on-one meetings with potential proposers, circulation of draft RFPs, and issuance of requests for information. State or local procurement law or practice may be more restrictive than the FAR with respect to information exchanges between an agency and prospective proposers.

In any event, procuring agencies must scrupulously avoid allowing any perception that one or more players (whether design or construction firms) has an inside track or greater access to material information. This is especially essential as construction industry players gain more experience with best-value selection methods, where contracts are not always awarded to the lowest qualified bidder. Many factors go into a firm's decision whether to compete for a particular project, and, as noted, the cost of participating in the procurement for a major subsurface project is high. To obtain maximum competition, it is critical that all channels, official and unofficial, deliver the message that it will be a fair contest.

In major projects of any type, firms often form joint ventures to submit a proposal, and experienced procurement officials understand that it takes time to find joint venture partners and reach agreement on basic terms of the relationship. Team building is particularly important in Design-Build projects. Many subsurface Contractors have histories of working in joint ventures with other Contractors and have industry contacts and previous relationships to draw upon. Far fewer of these Contractors have been involved in joint ventures with design firms to perform major projects on a Design-Build basis; therefore, greater time and effort is necessary for these firms to identify possible partners, explore compatibility and experiences, and negotiate the terms of their business relationship. In many cases these activities begin well before commencement of the formal procurement process. Accordingly, a procuring agency's early information releases about a project should highlight that it will be contracted on a Design-Build basis.

Many long lead-time activities must begin prior to commencement of the formal procurement process. The extent of geotechnical information and reports to be furnished by the Owner are an important issue that must be carefully considered. (The subject is discussed in Chapters 8 and 9.) All subsurface exploration data, as well as any reports interpreting the data or defining baseline conditions, should be developed as early as possible so that all data and reports can be double-checked for inconsistencies, omissions, or other problems, and because the data are necessary in developing the preliminary design. Many project issues involving third parties, such as right-of-way acquisition, utility relocation, environmental compliance and mitigation, and other permits and approvals, must be considered in a different light under the Design-Build approach.

In Design-Bid-Build, the Owner controls and deals with most of these third-party issues in coordination with development of the final design for the project, and many of the issues directly affect or are affected by the final design. In Design-Build, where final design is developed by the Design-Builder, the Owner must carefully consider the extent to which responsibility for some of these issues should be placed on the Design-Builder.

Subsurface projects, particularly those located in urban areas, often involve certain third-party interfaces of an extent and complexity greater than in most building projects. For example, a tunnel or other subsurface project will typically require acquisition of subsurface property interests from, and protection of existing improvements owned or leased by, many third parties. Impact or mitigation agreements may be more complicated than in other types of projects. Many of these third-party issues have political dimensions that the Owner can deal with more effectively and has greater stake in the outcome than the Design-Builder. It may not be possible for the Owner to fully resolve many of these issues prior to development of the final design, and their resolution will therefore become a shared responsibility. In some cases it may be possible to fully delegate certain matters to the Design-Builder—for example, in situations where the principal requirement is coordination of design and construction. Where discretionary approvals or politically sensitive issues are involved, simple delegation to the Design-Builder may be risky. In such cases, these issues could be kept by the Owner while the preliminary design is being completed.

Experienced proposer teams will try to cover themselves through price premiums, allowances, or proposal qualifications and assumptions if they are required to assume responsibility for activities that are outside of their core expertise and involve inherent uncertainty. Even if contract language purports to shift responsibility to the Design-Builder, the Owner may find itself subject to delays or additional costs if the Design-Builder is unable to deal effectively with third-party issues for reasons beyond its control. In some cases it may be advisable to include alternatives or allowances in the contract to deal with third-party issues that cannot be resolved by the Owner in advance and that carry significant potential risk.

For example, utility relocations can be a source of problems. If arranging for local utility companies to relocate their facilities is a relatively straightforward and predictable process in the municipality(ies) where the project will be constructed, the Owner may conclude that it is safe to include all necessary relocation (whether performed by the Design-Builder or the utility company) in the Design-Builder's base scope of work. This approach gives the Design-Builder the incentive to develop the final design with a view to minimizing utility relocation and with the potential to save money. The Owner must consider, however, the likelihood that it will realize any savings and consider the impact of possible conflicts over utility design solutions among the Design-Builder, the Owner, utility companies, or affected third-party property owners.

If the scope or other requirements for utility relocations are ill-defined because of lack of information concerning existing utilities or other reasons, the Owner may consider establishing allowances and assuming the risk of this uncertainty, such as by accomplishing utility relocation under separate early contracts. This may require a conservative approach, foregoing the opportunity to reduce relocation requirements through adjustments in final design, but will eliminate uncertainty and the possibility of delay claims or other disputes.

The best approach will depend on the particular circumstances of each procurement. In all cases the Owner and its consultants must consider the appropriate allocation of responsibility for third-party issues as early as possible, so that Contractor-responsibility items are clearly defined in the solicitation documents and Owner-responsibility items can be managed so as not to delay the critical path of the project after control has shifted to the Design-Builder. The issues must be worked through conceptually before commencement of the formal procurement process. Although problems often cannot be fully resolved in advance, the Owner and its consultants must have a concrete plan for solutions in order to determine what requirements should be imposed on the proposers. Planning for procurement should budget sufficient time and effort for all these issues to be addressed. A formal procurement process should allow opportunities for proposers to ask questions and receive answers. It should also allow adequate time for proposal preparation after proposers have received all pertinent information.

Managing the project schedule by compressing the procurement process runs the risk that proposers will not be sufficiently comfortable with some aspects of their proposals when the proposal deadline arrives. This uncertainty can result in higher contingencies carried in proposed prices. Inadequate time can also result in technical proposals that are not fully "baked," making them more difficult to evaluate. These concerns are particularly important in a Design-Build subsurface project because of the need for the proposer to develop and clearly document sophisticated design concepts. On the other hand, an overly protracted procurement timeline not only results in unnecessary schedule extension but also may raise proposer concerns about the Owner's commitment to the project or result in loss of focus and continuity within proposer teams.

After the preliminary design phase is completed and accepted by the Owner, it is prudent to make information available, including presentations, to various contractor and engineering associations, noting the details of everything that is available. There should be a clear statement of the finished project and that the RFP will provide detailed legal and contractual criteria.

CONCLUSION AND RECOMMENDATIONS

These conclusions and recommendations represent the authors' personal opinions, based on their experience and their review and evaluation of generally known industry experience with the Design-Build project delivery system over the past six to seven years. While it is clear that the use of Design-Build is growing, it is also clear that getting the procurement process right can go a long way toward success. Owners are not doing anything new other than creatively combining the design and construction functions into a single contract. Because the Owner knows what it needs in a finished facility, it is up to the Owner to assemble a team that can completely describe that facility and give maximum flexibility to the Design-Build team to deliver it in an efficient and cost-effective manner. The focus must be on clearly defining all project uncertainties and then fairly allocating the resulting risk to the appropriate team players. It goes without saying that some risk will remain with the Owner.

Recommendation 5-1: The Owner's team and its consultants need to clearly define all key project technical and management issues, thus reducing all uncertainties to as low a level as possible. At a minimum, the design should be carried to at least the 30% level, with adequate drawings, design criteria, and special specifications to define the finished product, while leaving means and methods largely to the Design-Build team. The geotechnical aspects should be nearly complete to save time and to reduce uncertainty.

Recommendation 5-2: The Owner must be familiar with formal procurement methods and the public and legal policies on which they are based before undertaking a large, complex procurement such as a Design-Build subsurface project. Careful selection procedures must include planning and a detailed and thorough documentation of the selection process, which must be effective and defensive to demonstrate the technical basis and objectivity in selecting the Design-Build team.

Recommendation 5-3: The Owner needs to understand the balance of control, time, cost, and technical approach that is required for a Design-Build procurement. Trust in the Design-Build team is essential to get the desired result.

Recommendation 5-4: The Owner should limit the number of qualified proposers because of the potential risk and financial expense involved.

Recommendation 5-5: The Owner needs to evaluate and decide, and make clear in the RFP, what will be the selection methodology: best-value, lowest-price–technically-qualified, or low bid with basic qualifications.

Recommendation 5-6: The Owner must create a clear and concise but comprehensive RFP that defines all elements to be submitted. The Owner should to be clear about what it needs, what risk it will retain, how the selection process will unfold, and how the project will be managed and controlled. A pre-qualification step that focuses on getting bidders that have experience both with Design-Build and of working effectively together in the past on similar projects is essential.

Recommendation 5-7: The Owner's team must communicate clearly to the proposers the evaluation criteria and how each is valued. Past performance on similar projects; design and construction expertise; legal, bonding, and insurance standards; and history of work-zone safety are all important and must be valued and weighed in a fair and reasonable manner. It cannot be understated how important it is to make these issues clear in the RFP.

Recommendation 5-8: As a general rule, a single contract package is preferable. Multiple contracts will create interfaces that are notoriously difficult to manage and that will increase management costs significantly.

Recommendation 5-9: To bring out the best teams and proposals, the Owner needs to evaluate the benefits and equity of a stipend for unsuccessful proposers. An honorarium will encourage more proposers after the pre-qualification stage and encourage a more detailed knowledge base for evaluation of the final proposals. An Owner should state whether or not all nonselected proposers will receive a stipend, or just the top two or three.

Recommendation 5-10: A prudent Owner will create a specific set of Design-Build contract documents rather than modify ones used in Design-Bid-Build work. The Owner will have to supplement its in-house team with legal, technical, political, and economic specialists (experienced in Design-Build projects) to adequately cover all aspects of risk definition and allocation, warranties, limits of liability, QA/QC enforcement, and so forth.

Recommendation 5-11: The Owner must plan how to do the procurement early in the process. Objectives and constraints, especially in public projects, must be identified early and considered when planning an RFP for a subsurface Design-Build project. Legislation to allow the process may be necessary to go forward. Effective outreach to both the public and design and construction communities is also necessary.

BIBLIOGRAPHY

Clinger-Cohen Act of 1996. 41 U.S.C. p. 235.

Cushman, R.F., and Loulakis, M.C. 2001. *Design-Build Contracting Handbook*, 2nd ed. Gaithersburg, MD: Aspen Law & Business, Chapters 9 and 10.

Didier, K.A. 1998. Construction contracting and the new two-phase design-build selection procedures: balancing efficiency with full and open competition. *27 Public Law Journal 589.*

Nash, R.C., Jr., and Cibinic, J., Jr. 1998. *Formation of Government Contracts*, 2nd ed. Riverwoods, IL: CCH.

Nash, R.C., Jr., Cibinic, J., Jr., and O'Brien, K.R. 1999. *Competitive Negotiation: The Source Selection Process*, 2nd ed. Washington, DC : George Washington University, Law School, Government Contracts Program.

Robinson, R.A., Kucher, M.S., and Gildner, J.P. 2001. Levels of geotechnical input for design-build contracts. In *Proceedings: 2001 Rapid Excavation and Tunneling Conference,* San Diego, California, June 11–13, 2001. Littleton, CO: SME.

CHAPTER 6

Agreements: Owner–Design-Builder and Design-Builder–Engineer

David H. Corkum, Esq.
Partner, Donovan Hatem LLP, Boston, Mass.

Michael R. Kolloway, Esq.
Senior Vice President and General Counsel, Metcalf & Eddy, Wakefield, Mass.

INTRODUCTION

Once a decision has been made to deliver a major subsurface project via the Design-Build method, the task of developing and negotiating the various agreements that will define the roles and relationships of the parties begins. Presumably, for a major subsurface project, the Owner will be a public entity, or the project will involve public funding, and the Owner will retain an Owner's representative to assist in the project. Often, multiple teams of Engineers and Contractors will be competing for the project. Ultimately, one of these teams will be selected and will enter into an agreement with the Owner. Accordingly, the types of agreements this chapter will discuss include

- Owner–Owner's Engineering Consultant,
- Potential Engineer–potential builder,
- Agreement between each of the potential Design-Build teams and the Owner,
- Owner–successful Design-Build team,
- Engineer and the Design-Builder within the Design-Build team, and, finally,
- Design-Builder and its subconsultants and subcontractors.

There are many permutations on the foregoing agreements. An Owner may, for example, be sufficiently sophisticated and endowed with in-house resources that it can prepare its own conceptual design. On another project, an Owner may insist upon including an operational and maintenance component in the main Design-Build construction agreement or, alternately, in a separate agreement. Another Owner may want to retain control over certain architectural elements of public facilities, while providing for a Design-Bid-Build contract for the overall concept of the project. Each project requires careful and deliberate contemplation of the interlinking network of agreements that will define the relationships between the various parties.

As with the Design-Bid-Build process, the Owner will have the strongest influences in setting the arrangement for various contracts. These agreements represent memorialization of the risk allocation and management philosophies as discussed in Chapter 3. Not only will the Owner's risk allocation philosophy be articulated in its agreement with the Design-Builder, but this philosophy will also strongly influence the agreement between the Design-Builder and its Engineer. Even the most carefully thought out and meticulously drafted agreement, however, is no match for the atmosphere of trust and cooperation that comes with a successful project. Here, too, the Owner will set the tone for this atmosphere during negotiation and drafting of the agreements. An Owner's unrealistic and unfair allocation of risk, coupled with complex proposal requirements without benefit of a proposal stipend, does not signal a trusting and cooperative relationship between the Owner and the Design-Builder.

The language and various provisions of the agreements will come into place as a means of defining the duties and responsibilities of the parties as well as a means of managing the downside of the project in the event it gets into trouble.

The Owner's first decision is whether to draft its agreements from scratch or to rely on one of the various standard form agreements as a base for its contracts. There are, of course, advantages and disadvantages to both. In almost no instance for a major subsurface construction project would a cut-and-paste of the standard agreement suffice to define the rights and responsibilities of the various parties. However, the standard forms have two significant advantages: (1) they dovetail together nicely; and (2) Contractors and Engineers are familiar with these forms and with the terms and conditions, both stated and implied. Although it is true that a 200-page bullet-proof agreement may serve the Owner well in protecting its downside interest, it will not promote the spirit of cooperation necessary for a successful project and will probably result in prolonged negotiations between the parties.

Among the most critical provisions of the Design-Build agreements are those detailing

- Standard of care,
- Guarantees,
- Warranties (both explicit and implied),
- Duty to indemnify and limitation of liability,
- Bonuses for early completion, and
- Liquidated damages for late completion.

Engineers and Contractors have certain standard expectations with regard to these provisions. An Owner may tinker with them, but doing so may have ramifications far beyond the Owner's intent. However, on a Design-Build project, where the justification for adopting a Design-Build delivery technique is single-point responsibility, the motivation to tinker with the generally accepted and expected standard of care, warranty, and guarantee may be irresistible.

On a subsurface construction project, differing site condition (DSC) is the most important risk consideration for an Owner. In fact, the Owner's desire to eliminate this risk may be the impetus for driving it toward Design-Build delivery in the first place. Key considerations include

- Whether a DSC clause will be included as a part of the contract language;
- Whether the Design-Builder can rely on the site surface investigation provided by the Owner; and
- Which party is responsible for the site investigations—the Owner, the representative, or the Design-Builder.

If the Design-Builder is required to perform its own site investigations, then the final contract negotiations must await either the final outcome of the investigation and interpretation of discovered expected conditions or provide for some flexibility for adjusting the contract price, depending on the final determination of conditions.

Permitting and third-party considerations are strategically important for the Owner intending to deliver a major subsurface project. This is a particularly complex issue when the project is constructed in an urban environment. As with DSC, problems with permitting can easily introduce significant delays and additional costs to a project. Likewise, interfacing with abutting Owners, neighborhood groups, and so forth can affect the final cost and general satisfaction associated with the delivery of the project. Design-Bid-Build permitting is almost always the responsibility of the Owner, for in the typical Design-Bid-Build delivery, the Architect/Engineer responsible for the design has both the time and the resources to perform the permitting on behalf of the Owner. In the Design-Build scenario, however, time is limited, and the Owner may not have sufficient information to be able to complete the permitting. Thus, on at least some Design-Build projects, the permitting is more efficiently performed by the Design-Builder.

Also, as with any contract, it is important to understand how to extricate one's self from a Design-Build contract in the event of either an unresolvable dispute between the Owner and Design-Builder, unforeseen circumstances that may arise after the execution of the contract, or for the mere convenience of the parties. Although there is no reason to believe this process is significantly different for Design-Build than Design-Bid-Build, considerations such as ownership of the documents merit special attention for a Design-Build subsurface project.

DESIGN-BUILD AGREEMENTS AND CONSIDERATIONS

For an Owner that has decided to deliver a subsurface project utilizing the Design-Build method, the agreement between it and its Design-Builder is by far the most significant document that will be produced. It will ultimately define, describe, and allocate the risks and responsibilities of the two principal parties most likely to bear the consequence of unforeseen events or occurrences. This document, however, will not be the first one executed on the project, and although the Owner should have a philosophical framework of how risks will be allocated, that framework can be expected to evolve as the conceptual design and project constraints are developed. Prior to the execution of this agreement, a series of predecessor agreements may be executed that will establish the Owner's philosophical framework for the Design-Build process and for the subsequent agreements.

Such prior agreements may include memoranda of understanding between

- Owner and its Engineering Consultant,
- Owner and potential Design-Build teams,
- Design-Builder and its Engineer at the proposal stage,
- Engineer and the Contractor of the selected Design-Build team, and
- Principals and their subconsultants and subcontractors.

Each of these agreements will describe the obligations among the principal parties and will be used as a benchmark for subsequent definitive agreements. It is certainly conceivable that gaps in scope and inconsistent definitions in these preliminary agreements may lead to disputes over the final agreements.

Several industry groups have circulated families of documents for use on Design-Build projects: the Engineers Joint Contract Documents Committee (EJCDC), the Associated General Contractors of America (AGC of America or AGC), the American Institute of Architects (AIA), and, finally, the Design-Build Institute of America (DBIA). The advantages of using one of these forms on a Design-Build subsurface project are consistency of terminology from one document to the next and the ability of such documents to address issues pertaining to the entire scope of the project. Certain presumptions, however, are built into the documents. For example, the responsible party for and the manner of conducting a geotechnical investigation, and the extent to which the Design-Builder may rely upon the findings of such an investigation, must be understood before a particular form of document is adopted. Choosing the wrong standard form and attempting to substantially modify it is probably not a wise move.

The DBIA booklet, "Design Build Contracting Guide," compares and contrasts various Design-Build issues of the common standard form documents (DBIA 1997). This booklet is an excellent source for an Owner attempting to evaluate the merits of the families of documents. As might be expected, the AIA documents and EJCDC documents tend to favor and protect the design professional's position throughout the documents, while the AGC and DBIA are more Contractor-friendly. The similarities and the differences between these standard form documents must be understood when considering their use for a Design-Build subsurface project. Thus, the following sections compare the common provisions of the agreements between an Owner and its Design-Builder.

American Institute of Architects

The AIA document pertaining to Owner and Design-Builder agreements is A191—Standard Form of Agreement Between Owner and Design/Builder, which comprises two parts (AIA 1996). The Part I Agreement pertains to preliminary design, budget, and schedule services required to accomplish the preparation and submission of the design/proposal, and the Part II Agreement covers services for final design and construction.

With respect to responsibility for site investigations, A191 requires that a Design-Builder "shall visit the site, become familiar with the local conditions, and correlate observable conditions with the requirements of the Owner's program, schedule and budget" (Part I Article 1, subparagraph 1.3.2). Although the agreement requires that the Design-Builder visit and become familiar with the site, the agreement does not require that the Design-Builder perform an exhaustive site investigation.

The document states that "services, information, surveys and reports required [in other paragraphs] which are within the Owner's control shall be furnished at the Owner's expense, and the Design/Builder shall be entitled to rely upon the accuracy and completeness thereof, except to the extent the Owner advises the Design/Builder to the contrary in writing" (Part II, Article 2, paragraph 2.7).

A191 includes a section to address DSC (Part II, Article 8, paragraph 8.5—Concealed Conditions, subparagraph 8.5.1):

> If conditions are encountered at the site which are (1) subsurface or otherwise concealed physical conditions which differ materially from those indicated in the Contract Documents, or (2) unknown physical conditions of an unusual nature which differ materially from those ordinarily found to exist and generally recognized as inherent in construction activities of the character provided for in the Contract Documents, then notice by the observing party shall be given to the other party promptly before conditions are disturbed and in no event later than 21 days after first observance of the conditions. The Contract Sum shall be equitably adjusted for such concealed or unknown conditions by Change Order upon claim by either party made within 21 days after the claimant becomes aware of the conditions.

Engineers Joint Contract Documents Committee

The EJCDC has developed three documents specific to Owner and Design/Builder agreements: Standard General Conditions of the Contract between Owner and Design/Builder (No. 1910-40) (EJCDC 1995a); Standard Form of Agreement between Owner and Designer/Builder on the Basis of a Stipulated Price (No. 1910-40A) (EJCDC 1995b); and Standard Form of Agreement between Owner and Designer/Builder on the Basis of Cost-Plus (No. 1910-40B) (EJCDC 1995c). No. 1910-40 works in conjunction with either document 1910-40A or 1910-40B. The documents are published jointly by the American Council of Engineering Companies, the National Society of Professional Engineers, and the American Society of Civil Engineers.

In addressing the responsibility for site investigation, document 1910-40A (Article 7, subparagraph 7.01) and 1910-40B (Article 11, subparagraph 11.01) requires the Design-Builder to represent that

> B. DESIGN/BUILDER has visited the Site and become familiar with and is satisfied as to the general, local and Site conditions that may affect cost, progress, performance or furnishing of the Work.

> D. DESIGN/BUILDER has carefully studied all reports of explorations and tests of subsurface conditions at or contiguous to the Site and all drawings of physical conditions in or relating to existing surface or subsurface structures at or contiguous to the Site which have been identified or made available by OWNER.

The EJCDC and AIA documents are similar in that neither requires the Design-Builder to perform an exhaustive site investigation. The difference between the documents is that the AIA document states that the Design-Builder "shall visit the site, become familiar with the local conditions, and correlate observable conditions with the requirements of the Owner's program, schedule and budget," while the EJCDC documents require the Design-Builder to make a representation that it has already performed the investigative work.

Like the AIA documents, EJCDC No. 1910-40 (Article 8, subparagraph 8.01, A, 6) requires the Owner to furnish information that the Design-Builder can rely upon in the performance of the Design-Builder's services. Specifically, the Owner shall:

> Furnish to DESIGN/BUILDER, as required for performance of DESIGN/BUILDER's Services the following, all of which DESIGN/BUILDER may use and rely upon in performing Services under this agreement:

> a. Environmental assessment and impact statements;

> b. Property, boundary, easement, right-of-way, topographic and utility surveys;

> c. Property descriptions;

> d. Zoning, deed and other land use restrictions;

> e. Engineering surveys to establish reference points for design and construction which in OWNER's judgment are necessary to enable Design/Builder to proceed with the Work;

> f. Assistance in filing documents required to obtain necessary approvals of governmental authorities having jurisdiction over the project;

> g. Subsurface data used in preparation of the Conceptual Documents.

EJCDC document 1910-40 includes a section pertaining to DSC (Article 4, subparagraph 4.02, A). The language (sub-subparagraph (i) of 4.02, A) is similar to the AIA concealed conditions language except while the AIA refers to prompt written notice of "subsurface or otherwise concealed physical conditions," the EJCDC's notice is for "subsurface or latent physical conditions." Sub-subparagraph (ii) is the same as the AIA document, which calls for prompt written notice to the Owner of "unknown physical conditions at the Site, of an unusual nature, which differ materially from those ordinarily encountered and generally recognized as inhering in work of the character called for by the Contract Documents."

Like the terms of the AIA agreement, upon notice of the DSC, the EJCDC agreement states that the Owner investigates the condition, and if there is a material difference in the conditions causing the Design-Builder's cost or schedule to change, an adjustment is made in writing by a change order.

Design-Build Institute of America

The DBIA has four documents that pertain to an agreement between an Owner and Design-Builder. The Standard Form of Preliminary Agreement Between Owner and Design-Builder (No. 520) is a stand-alone document containing its own general conditions and is used for preliminary services only, not for construction services (DBIA 1998). It should be used when an Owner decides not to contract for the complete design and construction at one time. After the preliminary document is executed, or if the preliminary document is not used, either the Standard Form of Agreement Between Owner and Design-Builder — Lump Sum (No. 525) (DBIA 2009a) or the Standard Form of Agreement Between Owner and Design-Builder — Cost Plus Fee with an Option for a Guaranteed Maximum Price (No. 530) (DBIA 2009b) can be used in conjunction with the Standard Form of General Conditions of Contract Between Owner and Design-Builder (No. 535) (DBIA 2009c). The DBIA documents are silent on the issue of responsibility for site investigation.

Included in No. 520 is language that gives the Design-Builder the right to rely on Owner-furnished information (Article 3, paragraph 3.3). Specifically, the Owner is required to provide

at its own cost and expense, for Design-Builder's information and use, the following, all of which Design-Builder is entitled to rely upon in performing its obligations hereunder:

1. Surveys describing the property, boundaries, topography and reference points for use during construction, including existing service and utility lines;

2. Geotechnical studies describing subsurface conditions, and other surveys describing other latent or concealed physical conditions at the Site;

3. Temporary and permanent easements, zoning and other requirements and encumbrances affecting land use, or necessary to permit the proper design and construction of the Project;

4. A legal description of the Site;

5. To the extent available, as-built and record drawings of any existing structures at the site;

6. To the extent available, environmental studies, reports and impact statements describing the environmental conditions, including Hazardous Conditions, in existence at the Site.

In addition, No. 535 includes language identical to the DBIA No. 520 preliminary agreement addressing the Design-Builder's right to rely on Owner-furnished information (Article 3, paragraph 3.2.1).

In addressing DSC, the DBIA's clause (Article 4, subparagraph 4.2.1) mirrors the DSC clause in the EJCDC's document No. 1910-40. Subparagraph 4.2.1 reads:

> Concealed or latent physical conditions or subsurface conditions at the Site that (i) materially differ from the conditions indicated in the Contract Documents or (ii) are of an unusual nature, differing materially from the conditions ordinarily encountered and generally recognized as inherent in the Work are collectively referred to herein as "Differing Site Conditions." If Design-Builder encounters a Differing Site Condition, Design-Builder will be entitled to an adjustment in the Contract Price and/or Contract Time(s) to the extent Design-Builder's cost and/or time of performance are adversely impacted by the Differing Site Condition.

Like the EJCDC and AIA agreements, the DBIA agreement requires that the Design-Builder promptly notify the Owner in writing of any differing site conditions.

Associated General Contractors of America

The AGC has assembled three documents that pertain to an agreement between an Owner and Design-Builder: (1) Preliminary Design-Build Agreement Between Owner and Design-Builder (AGC Document No. 400) (AGC 2007); (2) Standard Form of Design-Build Agreement and General Conditions Between Owner and Contracts (Where the Basis of Payment Is a Lump Sum) (AGC Document No. 415) (AGC 1993a); and (3) Standard Form of Design-Build Agreement and General Conditions Between Owner and Contractor (Cost When the Basis of Payment is the Actual Plus Fee with a Guaranteed Price Option) (AGC Document No. 410) (AGC 1993b).

The AGC Preliminary Design-Build Agreement is used in conjunction with No. 410 or 415, but its primary use is for the schematic design portion of a project only. If the parties intend to proceed beyond the schematic design phase, either document No. 410 or 415 would be used.

AGC document Nos. 410 and 415 are both silent in addressing which party bears the responsibility for site investigation. To a limited extent, No. 400 requires that the Design-Builder provide site investigation (Article 3, subparagraph 3.2.2). However, the investigation is limited to providing the Owner with a "preliminary evaluation of the site with regard to access, traffic, drainage, parking, building placement, and other considerations affecting the building."

Document Nos. 410 and 415 address whether the Design-Builder has the right to rely upon Owner-furnished site subsurface investigation information and opinions (Article 4, subparagraphs 4.1.2 and 4.1.4). In the pertinent part, subparagraph 4.1.2 requires that the Owner provide to the Design-Builder "all available information describing the physical characteristics of the site, including surveys, site evaluations, legal descriptions, existing conditions, subsurface and environmental studies, reports and investigations." Further, in subparagraph 4.1.4, "[t]he Design-Builder shall be entitled to rely on the completeness and accuracy of the information and services required by this Paragraph 4.1."

The AGC and AIA agreements more specifically state what the Design-Builder is entitled to rely upon. Although the EJCDC and DBIA agreements provide that the Design-Builder is entitled to rely on the information furnished by the Owner, the AIA and AGC agreements specifically state that the Design-Builder is entitled to rely on the "completeness and accuracy" of the information and services.

The issue of DSC is addressed under the heading of "Unknown Conditions" in AGC document No. 415 (Article 8, subparagraph 8.5) and in AGC document No. 410 (Article 9, subparagraph 9.4). Both paragraphs set forth language that essentially mimics the language and intent of similar DBIA and EJCDC clauses. "Unknown Conditions" in AGC document No. 410 (Article 9, subparagraph 9.4) reads as follows:

> If in the performance of the Work the Design-Builder finds latent, concealed or subsur-face physical conditions which materially differ from the conditions the Design-Builder reasonably anticipated, or if physical conditions are materially different from those nor-mally encountered and generally recognized as inherent in the kind of work provided for in this agreement, then the GMP [guaranteed maximum price] estimated Cost of the Work, the Design-Builder's Fee, the Date of Substantial Completion and/or the Date of Final Completion, and if appropriate the compensation for Design Phase Services, shall be equitably adjusted by Change Order within a reasonable time after the conditions are first observed. The Design-Builder shall provide the Owner with written notice within the time period set forth in Paragraph 9.6 [21 days].

In addition, AGC document No. 415 (Article 8, subparagraph 8.5) contains similar language, the exception being that the only differing item is the payment agreement.

The descriptions of standard form documents relate only to how they treat the subsurface and geotechnical aspects of a project in the Owner–Design-Builder agreement. As can be imag-ined, each of the standard form documents tends to promote the underlying philosophy of the promulgating organization. Thus, for example, the AIA and the EJCDC tend to be more sym-pathetic to the Engineer's position than is the AGC's document, which tends to lean toward the Contractor. Additionally, the documents vary in their approach to the method of selecting the Design-Builder and the presumption of whether the Design-Builder is led by a Contractor or Engineer. Thus, the first step in selecting among the standard form documents is for the drafter to determine that nothing is globally contradictory between the document's format and the project's procurement scheme.

If an Owner chooses to structure its procurement of a Design-Build subsurface project around one of these standard form document families, it will need to perform significant revi-sions to those sections addressing geotechnical data and baseline reports. These revisions will need to be surgical, precisely reflecting the Owner's intention. The Owner will need to resist the temptation to cut and paste what it sees as the best provisions from the individual standard form documents into an amalgamated composite. Doing so would almost certainly destroy the care-fully considered and coordinated *nesting* that is the very hallmark of these standard documents. If the overall structure and framework of these documents are unacceptable to the drafter, then starting from scratch or even attempting to modify a familiar and otherwise comfortable Design-Bid-Build agreement probably would be preferable rather than confusing readers with one that looks like a standard form document but has been radically altered.

AGREEMENT BETWEEN OWNER AND OWNER'S ENGINEERING CONSULTANT

Well before the substance of the Owner/Design-Builder agreement is drafted, the Owner of a major subsurface project will have entered into an agreement with its Engineering Consultant. The typical scope of services performed by the Owner's Engineering Consultant includes provid-ing a conceptual design, developing design criteria, assisting in the procurement and selection of a Design-Builder, and monitoring the project on behalf of the Owner during construction. The agreement between the Owner and its Engineering Consultant will most likely resemble the standard professional services agreement that Owners and Engineers typically enter into for traditional Design-Bid-Build projects.

The American Consulting Engineer's Council, in its recently published booklet, "Design-Build Project Delivery," recommends the standard form document developed by the EJCDC, No. 1910-43 (EJCDC 1995d). Other standard form contracts are available, and many public Owners will undoubtedly have their own standardized agreements. No matter which form is used, both parties for a major subsurface project should pay careful attention to contractual provisions that upset or reverse general expectations that Owners, Engineers, and Contractors have come to believe are cast in stone.

As with Design-Bid-Build projects, the fundamental purpose of this agreement is to articulate the scope of services and deliverables that the Engineer will perform and provide and to establish the terms of compensation. The standard form agreements tend to discuss the scope of services in phases that temporally track the development of the project—for example, conceptual design, proposal evaluation, and construction support. The need for a well-thought-out, fully discussed, and negotiated scope of services is obvious.

Equally important is the need to define which party will perform the services. Owners often insist on the commitment of key personnel who are familiar with the Design-Build process for the duration of the project. The Owner's leverage in insisting on the commitment of certain individuals within the Engineer's organization is strongest during the contract negotiations and progressively dissipates. The Owner may find it difficult to retain an Engineer's key personnel for additions or modifications to the original scope of services due to the commitment of such personnel to other projects.

Agreement Considerations for Developing Design Criteria

Single-point responsibility, as has been mentioned many times in this book and has generally come to be accepted in the industry, is one of the prime motivations for adopting a Design-Build project delivery mechanism. When the final product of a Design-Build project fails to meet the specified contractual objectives, the Owner need only look to the Design-Builder to remedy the problems, whether such problems result from defective design or workmanship. What happens, however, if the design criteria are somehow defective? In that case, if the Design-Builder is exculpated, can the Owner look to its Engineering Consultant for a remedy?

Conceptually, the general allocation of responsibility for Design-Build project performance is quite similar to the allocation for temporary works on a Design-Bid-Build project. Excavation support, for example, on a subsurface Design-Bid-Build project will most likely require the Contractor to design the system to achieve certain Owner-specified performance criteria related to surface settlement. If the Contractor fails to achieve the performance criteria, then, barring other mitigating circumstances, the Design-Builder will probably be contractually obligated to remedy the situation at no cost to the Owner. In the same manner on a Design-Build project, the Design-Builder will be held responsible for obtaining the performance criteria of the final facility. What happens if, despite the ground being exactly as depicted, and the Design-Builder performing its work flawlessly, the facility exceeds the performance criteria?

Consider, for example, a specified infiltration–exfiltration criterion for a deep-rock combined sewage overflow system project. Unbeknown to any of the project participants, the joint mineralization, which contributed to the tightness of the rock and led to the stated presumption that an unlined tunnel could meet the criterion, is partly composed of a calcareous material. After a few years of operation and exposure to an acidic environment, the joint filling has begun to deteriorate, and the infiltration criterion is no longer met. Which party, the Owner or the Design-Builder, will bear the responsibility for remedying the problem? In either case, can the disappointed party then turn to the Owner's Engineering Consultant for compensation? The answer to these questions is, of course, "it depends." It depends on the explicit terms of the various agreements and whether the Owner's Engineering Consultant has agreed to bear that risk.

It is worthwhile to reiterate at this point that many scholars and practitioners could agree on who *should* bear the risk in this situation. To arrive at that determination, they would simply ask which party was in the best position to control the risk. If, for example, the Owner's Engineering Consultant was solely responsible for the entire geotechnical investigation, was unencumbered by time or budget considerations in performing that investigation, and produced all the relevant baseline reports, then it would be the prime candidate for the fair allocation of that risk. If, on the other hand, the Engineering Consultant's geotechnical investigation was only at reconnaissance level and was delivered with appropriate caveats to the Design-Builder for finalization consistent with the intended means and methods, then it would be appropriate to assign the risk to the Design-Builder. Finally, if the Owner had limited resources available to its Engineering Consultant to adequately perform an investigation, then the risk of error should be borne by the Owner if such restrictions impacted the Engineering Consultant's ability to conduct a complete and proper investigation.

Notwithstanding the common notion of who *should* bear the risk, which party actually bears it will most likely be determined by the terms of the agreements. A clear, conspicuous communication in the agreement between the Owner and its Engineering Consultant that the risk of an unobtainable design criterion was assigned to the Owner could insulate the Engineering Consultant. However, if the Owner has allowed its Engineering Consultant reasonable autonomy to perform the investigation, it seems unlikely that the Owner would agree to assume this role. Both the Owner and its Engineer should carefully evaluate the design criteria that will be provided to the Design-Builder and consider which party is in a better position to manage the risk. Based on this evaluation, the responsibility can be negotiated and insurance coverage may be obtained to cover the risk.

Agreement Considerations for Geotechnical Investigations

On all major subsurface projects, the Owner's Engineering Consultant will perform some level of geotechnical investigation. The Engineer may perform the geotechnical investigation with its own forces or retain a geotechnical subconsultant, in which case the Contractor will attempt to *flow down* its obligations to the Owner for the investigation to its subcontractors. Chapter 8 discusses the pertinent considerations of which entity—the Owner, the Owner's Engineering Consultant, or the Design-Builder—should perform the geotechnical investigation. Chapter 9 addresses the recommended content of the reports as well as their status as contract documents. The arrangements have dozens of permutations, and the contract between the Owner and its Engineering Consultant must fairly and adequately address the role and responsibility of the Owner's Engineering Consultant.

In one permutation, for example, the Owner might expect its Engineer to develop a 30% conceptual design, including geotechnical investigation and reports adequate to support the procurement of a Design-Builder for the project at a fixed price. The well-known Spearin doctrine, which imposes an implied warranty by the Owner to the Contractor that the project, as described, can be built in the subsurface conditions as indicated, is as applicable in Design-Build projects as Design-Bid-Build. Thus, if some aspect of the conceptual design or design criteria is faithfully executed by the Contractor, but the resultant product is impossible to achieve or is unsatisfactory due to a lack of compatibility with the ground conditions, the Owner will bear that risk. The question then becomes: Can the Owner look to its Engineering Consultant to recoup its loss?

In another permutation, consider the same circumstances, but the Contractor's failure to perform is the result of differing ground conditions from the conditions originally indicated. In this case, the issue should be addressed in the Owner–Contractor agreement. If the Owner bears the loss, can it turn to its Engineering Consultant for reimbursement? If, on the other hand, the Contractor bears the loss, is there a mechanism by which it could recover against the Owner's Engineering Consultant for a defective geotechnical investigation or report?

As a final example, consider the case where the Owner's Engineering Consultant's geotechnical investigation and report was not meant to be the final and definitive indication of subsurface conditions but only a starting point for the Design-Builder to develop its baseline. In the event that errors or omissions in the Owner's Engineering Consultant's geotechnical investigation/ report lead to a loss by the Design-Builder, the rights and responsibilities between the Owner and Design-Builder will be articulated in their agreement. What, however, are the rights of recovery by the disappointed party against the Owner's Engineering Consultant? Although the Owner will obviously be entitled to seek direct contractual remedies against its Engineering Consultant, in order for the Design-Builder to do so, the agreement between the Owner and the Design-Builder must specifically address the latter's ability to rely upon the Engineering Consultant's report, and either: (1) insulate the Design-Builder from liability as a result of its reliance upon the report, or (2) specifically describe the Design-Builder's rights against the Engineering Consultant. If the agreement between the Owner and the Contractor provides certain rights by the Design-Builder against the Engineering Consultant, then the Owner's agreement with the Engineering Consultant must also reference and describe such rights. (For a comprehensive review of the risk allocation principles related to performing and providing geotechnical information on a Design-Build project, see Hatem 1998, Section 10.2.3 of Chapter 10, pp. 341–358.)

Alternate Concept: Owner's Independent Geotechnical Engineer. It is conceivable that an Owner contemplating a Design-Build subsurface project might choose to enter into a contract with a geotechnical consultant separate from its Engineering Consultant. Possible motivating factors include a preference for that particular geotechnical consultant based on expertise in the area; a complex geologic regime coupled with difficult access and logistical constraints that necessitate beginning the geotechnical investigation before the Owner's Engineering Consultant can be selected; or a desire by the Owner to have continuity in its geotechnical consultant but not necessarily in its Engineering Consultant throughout the project.

If the Owner retains an independent geotechnical consultant, the allocation of risk between the Owner and its Engineering Consultant is greatly simplified, and their agreement should explicitly state that the Engineering Consultant has no responsibility, and thus no liability, for geotechnical-related problems. In this hypothetical situation, the more the Engineering Consultant attempts to direct and influence the activities of the independent geotechnical consultant, the greater the opportunity for its entanglement in that liability.

Agreement Considerations for Permitting

Permitting issues, just like the geotechnical program for a project, can also become the source of large and costly delays, and, just like the geotechnical program, numerous configurations of scope and responsibility can be assigned to the Owner's Engineering Consultant. The two extremes range from situations where the Engineering Consultant assumes complete responsibility for obtaining all required permits (with the exception of those obtainable only by the Contractor) to situations where it simply identifies the permits that the Design-Builder must obtain. The primary consideration for drafting or negotiating an agreement is to clearly define the scope of services in order to adequately address these issues.

In the event the Owner's Engineering Consultant is fully and completely responsible for permitting, then its activities could very well end up on the critical path of the project. In such a situation, the Owner would most likely compensate the Design-Builder's for delays. If so, under what circumstances can the Owner look to its Engineering Consultant for recovery of those costs? If the Contractor is prohibited from seeking reimbursement from the Owner in the case of delayed permitting, then its contract must provide for an extension of time if permits are delayed.

Whether the Owner's Engineering Consultant is fully responsible for the permitting or is simply beginning the process and passing it off to the Design-Builder, there will be a timely cooperative sharing of information to complete the process. If the Engineering Consultant is fully responsible and financially liable for delays associated with untimely permit issuance, then it must ensure that it is able to require the timely and accurate supply of information.

For example, a discharge permit to allow dewatering of an excavation might require that the Design-Builder commit to means and methods of ground support earlier than desired. The permits will be based upon the selected technique and associated estimates of quantity and quality of discharge. If the Design-Builder subsequently determines that it is appropriate to implement a different system, the permit may be insufficient for this system, and the project may be delayed as the new permits are obtained. Under this scenario, it is important that the agreement between the Owner and the Design-Builder address this possibility and the issues of delays and additional costs associated therewith.

Agreement Considerations for Evaluating Proposals

On a major Design-Build subsurface project, the Owner's Engineering Consultant will most likely have drafted, or at least assisted in developing, the procurement documents by which the Owner solicits proposals for potential Design-Builders. Likewise, it will probably assist the Owner in evaluating those proposals and recommending a team to the Owner.

Agreement Considerations During Construction

Once the Owner has engaged a Design-Builder, the Owner's Engineering Consultant's role will and should take a back seat to that of the Contractor. Although the Engineering Consultant's role will be reduced, it still has an important function in the project's successful completion. The agreement between the Owner and its Engineering Consultant must recognize this shift in roles and address the Engineering Consultant's ability to influence the project and the limitations to its authority, and specifically describe this role in the written scope of services set forth in the Engineering Consultant's agreement.

In many respects, the Engineering Consultant's role during the detailed design/construction phase of the project will be familiar and natural to Engineers currently performing on Design-Bid-Build subsurface projects. Specifically, refer to the role the Design-Bid-Build Engineer adopts when reviewing, for example, a Contractor's proposed support of excavation. In this example, the Engineer reviews submittals for conformance with specified performance criteria, such as water tightness or surface settlement. Similarly, in the Design-Build situation, the Owner's Engineering Consultant will be looking for conformance with performance criteria for the finished product, not just for the temporary works. The aggressiveness of the Engineering Consultant in reviewing the Contractor's submittals should be proportional to the criticality of that submittal's contribution to project performance. If, for example, the Engineering Consultant is expressly warranting in its agreement with the Owner that a performance criterion is achievable—for example, surface settlement and impact to abutters when using the New Austrian Tunneling Method as a subsurface excavation technique—it would be well advised not to try and completely abdicate its role in the submittal review.

If the Owner–Engineering Consultant agreement provides that the Engineering Consultant will take the lead in performing geotechnical investigations and permitting services during the design process, then careful consideration needs to be paid to the responsibility for project delays associated with delays in permitting and obtaining additional information. If completion of the project is tied to a source of revenue, then the Owner may attempt to pass the cost of the lost revenues on to its Engineering Consultant in this case. If so, the Engineering Consultant should be certain that it is in a position to control this risk and that it is being adequately compensated to assume this risk.

There are other ways that the inadequate performance of the Owner's Engineering Consultant could introduce a costly delay to the project. The most obvious example is delays in reviewing and addressing the Design-Builder's submittals. The Owner–Design-Builder agreement will undoubtedly delineate a time frame for response to these submittals. Failure on the part of the Engineer to adhere to the window will result in a schedule extension and perhaps damages to the Design-Builder. The agreement must thus define whether the Owner absorbs that time and costs or they are passed along to the Engineer.

In the event that the Design-Builder encounters a claimed DSC during the course of the work, which delays the project, then the Owner will most likely request that its Engineering Consultant assess the merits of that DSC claim and associated damages. In order to do this in a meaningful way, the Engineering Consultant will need to have a regular presence on the site and be completely familiar with (without assuming any responsibility for) the Design-Builder's means and methods of construction and associated performance. If the agreement requires the Engineering Consultant to make this type of assessment, then it also should anticipate its commitment of appropriate resources.

AGREEMENT BETWEEN OWNER'S ENGINEERING CONSULTANT AND ITS SUBCONSULTANTS

Barring a specific requirement in the Owner–Owner's Engineering Consultant agreement, the Engineering Consultant is free to contract in any manner it desires with its subconsultants. Often, an Owner may request the right to approve subcontractors for the project; however, such approval must usually be in the Owner's reasonable discretion. The Owner's Engineering Consultant could, however, be faced with unexpected exposure to damages if it fails to adequately coordinate key provisions of its agreement with its subconsultants with similar provisions of its agreement with the Owner. Moreover, in those situations where the subconsultant plays a major role in the project, such as a geotechnical or tunnel design firm, there are some additional considerations.

The best example of failure to adequately coordinate can be found in a mismatched standard of care. For example, if the Owner–Engineering Consultant agreement required that the Engineering Consultant provide its services using the "utmost care," and the agreement between the Engineering Consultant and its subconsultant is either silent on the standard of care or provides for the subcontractor to perform its services in accordance with an "industry standard" or "acceptable" standard of care, the Engineering Consultant will be held to a higher standard or level of responsibility than its subconsultant. Thus, if an error or omission has been detected in a performance criterion developed by the subconsultant and incorporated into the conceptual design by the Engineer, and if the project does not perform as required, the Owner may seek reimbursement from its Engineering Consultant. If the Engineering Consultant now turns to its subconsultant for a contribution to the loss, it may be disappointed to learn that it has promised more to the Owner than its subconsultant has promised to it. In the absence of a specific provision in an agreement, the common law of all states will impose a duty of reasonable care on the subconsultant. If the loss occurred because of a failure of the subconsultant to perform its services somewhere between its standard of reasonable care and the Engineering Consultant's higher standard of utmost care, then the Engineering Consultant will have to absorb the loss.

Another simple example of coordinating contract provisions is a pay-when-paid or a pay-if-paid clause. If the Owner has the ability to withhold payment to its Engineering Consultant because of inferior or unsatisfactory services performed by the subconsultant, then the Engineer may want to insulate itself from becoming a surety for its subconsultant by including a provision that the Engineering Consultant is not required to remit payment to the subconsultant for its invoices until, or if, it receives payment from the Owner.

Although not unique to Design-Build projects, where the subconsultant is a key player in the overall project, the parties may prudently consider naming the specific individuals in the agreement and assuring their continued involvement in the project.

AGREEMENT BETWEEN OWNER AND POTENTIAL DESIGN-BUILDER

The simple rule of supply and demand suggests that an Owner of a subsurface project will get a better price if there are multiple, willing Contractors competing for a contract. Likewise, with Design-Build projects, an Owner will be in a better position to insist upon predetermined terms and conditions if it has multiple Design-Build teams competing for the project. There is certainly a point of diminishing returns where too many proposals could overwhelm the Owner's ability to responsibly evaluate them. On major projects, this is seldom a problem due to the limited pool of capable Contractors and Engineers and the tendency to pool resources.

If multiple Owners in different parts of the country are tendering projects at the same time, Contractors are going to focus upon the more desirable projects. This may affect the available pool of Contractors for smaller or higher-risk projects. Not only are the resources to prepare the bids limited, but the costs of doing so are also often significant. Thus, Contractors allocate their business development dollars to those projects they perceive have the best likelihood of success. Preparing a Design-Build proposal is generally thought to be at least twice as costly as preparing a simple bid on a Design-Bid-Build project. How does an Owner assure itself a field of qualified potential Design-Builders?

Timing is perhaps the most important and least controllable factor. A more utilitarian approach to attracting proposals is to help defray the cost of preparing the proposals. This is done by providing a stipend to those Design-Builders invited to propose on a project. The amount of the stipend typically ranges from the estimated amount required to prepare the entire proposal to the presumed difference between the cost required to prepare the Design-Build proposal and a more typical Design-Bid-Build bid. The objective is to attract teams, and the agreement with the prospective teams can be quite simple, assuming that a public Owner is not encumbered by an onerous procurement system.

A simple agreement is required between the Owner and qualified Design-Builders. The agreement need only specify the scope of the proposal to be delivered and the terms and timing of the stipend payment. If an Owner is concerned about the quality of the proposals, it has probably not done an adequate job in qualifying and preselecting the teams.

Public Owners will have to comply with their state laws and procurement regulations about preselection of proposal teams. A Design-Builder's prior successful experience with similar projects is bound to be a key prescreening criterion for which teams will be allowed to compete and be offered a stipend. Many Owners on Design-Bid-Build projects are unable or unwilling to prequalify bidders, sometimes rationalizing their action by stating that "the bonding companies will pre-qualify the contractors." A successful Design-Build subsurface project, however, requires more than the financial stability and cash flow potential of a major construction company. Although size and bonding capacity are important factors, the Owner should be equally interested in the team's demonstrated ability to innovate. Bill Gates, in response to a question about what could displace Microsoft's superiority, remarked that he was not concerned about the competition from other software companies. Instead, he was worried about the 15-year-old working on an original idea. Ideas and innovation come from people, not companies; therefore, an Owner's preselection should also heavily weigh the résumés of the key people proposed as team members.

During the preselection process, the Owner may require all bidders to identify any exceptions taken to the Owner's key proposed contractual terms, such as standard of care, indemnification, guarantees, or liquidated damages. Thus, the Owner may consider the Contractor's willingness, or lack thereof, to accept the Owner's proposed contractual terms during the preselection process.

The issue of confidentiality of the potential Design-Builder's proposal is often debated. Which party owns the design, the schedule, the ideas, and the cost estimates developed for a proposal? Can the Owner pick and choose the best concepts from the various proposals, disclose them to its preferred Contractor and then negotiate a contract price with that entity? Certainly, if the Owner provided a stipend large enough to cover the entire proposal production, it could convince itself that it was morally entitled to use this work-for-hire to its advantage. Such an intended or unintended (but possible) lack of confidentiality should be carefully considered by the Owner as it will tend to stifle the very innovation that the entire Design-Build process is attempting to foster. Moreover, the question of liability for the design concepts becomes a concern not only for the originator and the recipient of that concept but also for the Owner that has facilitated the exchange.

Notwithstanding those types of information that probably should be considered confidential, the agreement should explicitly state that factual information developed or discovered by a competing Design-Builder and paid for with an Owner's stipend will be shared with all parties. Subsurface data, for example, should belong to the Owner but not necessarily the interpretation of that data. Every team and the Owner will benefit from having as complete a picture as possible of the ground conditions. The competitive distinction should be based on how each Design-Builder chooses to interpret and utilize that data.

AGREEMENT BETWEEN OWNER AND DESIGN-BUILDER

The agreement between the Owner and its Design-Builder will probably be the single most important document that is drafted for the project. It will define and describe the rights, liabilities, and obligations between the two parties. As discussed at the beginning of this chapter, the owner may utilize one of the standard form documents or draft one from scratch. In either case, the Owner will have the benefit of input from the Owner's Engineering Consultant. Additionally, the Owner may intend and provide for an extensive negotiation process with the selected Design-Builder prior to the agreement's execution.

The principles of risk allocation that the agreement will intend to articulate have been comprehensively described and treated in Chapter 3. In order to demonstrate the deliberation that should accompany the drafting of the agreement, this section discusses several of the major considerations for Design-Build subsurface projects.

For purposes of this discussion, we will summarize Chapter 3 with the following simplified maxims:

- The risk should generally be assigned to the party in the best position to control that risk; and

- The financial benefits to a party should conform to the risks being assumed.

Perhaps the single most significant risk on a subsurface construction is whether or not the actually encountered ground conditions and behavior conform to the model and assumptions stated at the time the contract price was established. Close behind differing subsurface conditions, (although in no way unique to subsurface projects) is the risk of interference from third parties. Finally, the risk of utility relocations and removing other suspected but not adequately defined obstructions has been the source of many major disputes on subsurface projects.

Differing Site Conditions

Nearly all major Design-Bid-Build subsurface projects in recent history have included a DSC clause. Federally funded contracts during the previous 75 years have included the language prescribed in the Federal Acquisition Rules, 48 CFR §52.236-2(a). Most state procurement laws require similar language in their state-funded projects. The purpose of this contractual provision is to allow a Contractor to be compensated for unanticipated costs incurred as a result of ground conditions that differed from that indicated or was reasonably inferred from the contract documents.

The subsurface construction industry universally supports, promotes, and has come to expect that the clause will be included in contracts. It allows Contractors to estimate and bid work based on what the Owner has told them to expect, with the assurance that if conditions change and costs increase, an adjustment can be made. Accordingly, the Contractor does not include a contingency in its bid for the possibility of performing the work under less favorable conditions, and (or so the theory goes) the Owner pays for the real value of the work. As anyone in the industry will attest, it doesn't always work that way. Stubborn Owners and greedy Contractors make for disputes. The industry, however, seems to have accepted those disputes as a tolerable part of the marketplace's friction. An Owner's reputation for generating abnormal amounts of friction may result in higher costs and fewer bidders on a given project, and a Contractor's similar reputation can draw a less accommodating Owner. Knowing what we do about DSC clauses on Design-Bid-Build projects, how should a Design-Build subsurface project's agreement deal with the issue?

Where the Owner or its Engineering Consultant has performed all the geotechnical investigations and collaborated with the Design-Builder to provide interpretations, predict ground behavior, and establish baselines, there is no reason that the Design-Bid-Build model for resolving DSC should not be adopted. In this situation, the Owner has controlled the investigation and had the opportunity to develop an appropriate baseline. Through its Engineer, it can adopt an aggressive or conservative approach to establishing that baseline. By working with the Design-Builder, it gains the benefit of knowing the Design-Builder's means and methods and can tailor the baseline to that approach. In this situation, if the ground condition differs, causing a cost increase, the Owner should expect to compensate the Design-Builder for those additional costs. If, on the other hand, the project is set up where the Design-Builder performs the majority of the geotechnical investigation and interpretations, authors the reports, and establishes the baseline, then the generally accepted notion of a DSC clause may not be appropriate. The Owner in this situation may very well assign the geotechnical investigation responsibilities to the Design-Builder for the purpose of absolving itself from the risk of a DSC claim.

Consider, for example, a soft ground tunnel project where the selected Design-Builder has determined that it could economically mine the tunnel using an open-faced tunnel boring machine (TBM). The Design-Builder completes the geotechnical investigations, confirms its assumed interpretation of standup time for the ground, and baselines an acceptable number commensurate with its proposed excavation techniques. How receptive do you think the Owner will be to a claim from the Design-Builder that it must now retool its tunneling plan to an earth pressure balance (EPB) machine because the ground is behaving differently than the Design-Builder assumed in its baseline report? And, just to make it interesting, let's say the other potential Design-Builders had all intended to use EPB machines from the start. In this situation, it is difficult to justify that the Owner should shoulder the additional costs for this DSC. The Design-Builder was responsible for and in control of both the investigation and interpretations. It provided the erroneous baseline of standup time that led to the wrong selection of equipment, and, according to the two global maxims at the beginning of this chapter, the Design-Builder should bear the burden.

In another example, consider a hard-rock tunnel project where, as in the previous example, the selected Design-Builder has performed the majority of the geotechnical investigation and interpretations, authors the reports, and establishes a baseline for the rock's unconfined compressive strength. The Design-Builder's cost estimate and its contract price are based on a productivity tied to the TBM's penetration rate as determined from the ground's unconfined compressive strength. Would the Owner be any more understanding if the ground turned out to be significantly stronger and thus more difficult and costly to penetrate? In this situation, the ground is truly different, both physically and mechanically, than the baseline report indicated. Unlike the former example, where the claim was somewhat attenuated by the Design-Builder's interpretation of an assumed ground behavior characteristic, the difference in this example is easily measurable and its effect more generally accepted. The agreement should clearly articulate which types of DSC claims the Owner is willing to entertain and what types are solely the Design-Builder's responsibility.

The simple truth is that some substantial risk will be associated with every subsurface project that in some shape or form involves differing subsurface conditions. The Owner should realize that if it attempts to completely disavow responsibility for any and all DSC based on the rationale that the Design-Builder controls the investigation interpretation and baseline process, it will surely pay a premium in the form of contract contingencies for the work. Several techniques attempt to minimize the costs of that risk.

One approach that has been used in the past is for the Owner to establish a contingency account sufficient to cover the costs of potential DSC claims (for example, 10% of the excavation costs). This account is used to fund all or certain pre-established categories of DSC claims. At the end of the project, any remaining funds are split equally between the Owner and the Design-Builder. This system removes the need for the Design-Builder to carry a contingency and provides an incentive for the Design-Builder to minimize its assertions of DSC claims in hopes of reaping a bonus at project completion. Another approach would be for the Owner to agree to pay all or certain pre-established categories of legitimate DSC claims but only after the Contractor had first absorbed an initial or deductible amount.

DSC are a matter of concern during the construction of the facility and for the viability of the facility's design. Unanticipated and unidentified in-situ stresses can overcome the facility's structural design. Unaccounted for aggressive groundwater or stray electrical currents can wreak havoc with the facility's corrosion resistance and design life. A site hydrology can be upset by preventing or promoting groundwater flow that can have both proximate and remote impacts to third parties at some future date. These are all examples of potential future problems, some involving third parties, that the Owner will need to evaluate and determine which party should bear the responsibility and how that risk is articulated. These types of risks implicate professional standard-of-care issues, Design-Builder's express and implied warranties, and the adequacy of the geotechnical investigation and validity of interpretations. (For an in-depth consideration of the interrelation of these issues of design adequacy and geotechnical investigation, please see Hatem 1998, Chapter 10, pp. 278–361.)

Third-Party Impacts

Permitting agencies, abutters, citizen groups, local fire departments, and adjacent contractors have all wreaked havoc on the schedule and budget of major subsurface projects. These third parties generally share the common characteristic of complete indifference to the contractual obligations between the Owner and its Design-Builder. There is no reason to believe that these impacts would be any different whether the project is delivered via Design-Bid-Build or Design-Build. Much of the public perception of any construction project is projected by the Contractor. If the Contractor is neat, orderly, and generally responsive to the needs of the community, then the relationship with the community is smoother. Noise, dust, nighttime blasting, and construction traffic are all under the control of the Design-Builder, thus making it responsible for community relations. The negative effect of failed or strained relations seems a rational extension of our two basic maxims. Other impacts, such as community input and signoff on the architectural features of the facility, are definitely not the types of risk a Design-Builder should be expected to absorb. Any changes to these features after the Design-Builder has provided a contract price should entitle it to a price adjustment.

Utility Relocation and Other Manmade Obstructions

A major subsurface project in an urban area is going to experience problems with relocating buried utilities. Short of shutting down the entire street and concentrating on nothing but utilities until they are completed, there is no efficient, unobtrusive way to perform this work. The only universal lessons learned from past jobs are: (1) get started early, and (2) if possible, have the work performed by a separate contractor. In many ways, a Design-Build subsurface project is well suited to utility relocation work simply because the Design-Builder can start immediately and relocate only those utilities needed for its selected construction techniques. On one recent Design-Build project, the contract budgeted an amount for utility relocation because it could not easily be defined in advance. The parties negotiated a conservative figure with the understanding that any amount not spent would be divided equally at the end of the project. This provided an incentive to both parties to try to find ways to minimize the cost of utility relocation.

Also in urban areas, especially older cities, subsurface construction projects are bound to encounter manmade obstructions, such as old piles, granite block foundations, retaining wall tendons, and abandoned utilities. Although the presence of these obstructions is known in general, specific information such as location, number, and character is usually lacking. On a Design-Bid-Build project, the Engineer often provides a contingent unit-price item for dealing with these obstructions and compensating the Contractor on either a per-each or volumetric basis. Contractors have a difficult time estimating the cost of dealing with these obstructions, no matter whether they are cleared on a wholesale basis in advance of the main construction effort or as the obstructions are encountered.

On a Design-Build project, the dilemma of determining how much effort to spend in identifying and characterizing these obstructions in conjunction with the cost of removal and the risk of unanticipated problems is simplified. Arguably, the Design-Builder can more efficiently and precisely focus its investigation effort because it is aware of the intended means and methods of construction. Nevertheless, there will still be a risk, albeit a theoretically smaller one, that must be accounted for, and the Design-Builder will carry a contingency amount. If the Owner chooses to assume this risk, for instance by providing a contingency fund similar to the utilities or DSC funds described previously, then it needs to thoroughly understand what constitutes an unanticipated obstruction and its impact on the project. The possibility of the Design-Builder manipulating the impact of the obstruction in order to draw from the contingency fund, coupled with the Owner's inferior position in trying to refute that impact, suggests that this risk provision requires careful, project-specific consideration and negotiation between the Owner and Design-Builder.

Schedule Milestones and Guarantees

Every Owner wants its project to be delivered on schedule. The driving force behind the schedule for subsurface projects can range from court-ordered milestones for environmental abatement projects, to coordination with an adjacent contractor's sequence of work, to facility startup on revenue-generating projects (such as cooling water tunnels for a power plant). Additionally, an Owner may insist upon performance milestones and may require that a Design-Builder guarantee a certain level of performance by the identified milestone dates. The Owner may attempt to assess liquidated damages if the schedule identified in the agreement is not satisfied. In such an event, a Design-Builder will attempt to include contractual language to provide that liquidated damages will not be assessed in the event of a failure to meet contractual milestones as a result of force majeure, unforeseen conditions, or DSC.

For those projects where time is the driving force of the contract, where significant liquidated damages are assessed to the Design-Builder for late performance, and where the Owner incurs commensurate or even greater costs as a result of late performance, time may be the Owner's contribution to a problem qualifying as force majeure, unforeseen conditions, or DSC. Under this arrangement, the Owner agrees to grant a time extension and thus relieve the Design-Builder of its obligation to pay liquidated damages. The Owner absorbs its loss from not having the facility when originally promised, and the Design-Builder absorbs the additional cost of remedying the underlying problem.

Hazardous Material CERCLA Indemnity

As established in the Comprehensive Environmental Response, Compensation, and Liability Act (CERCLA), commonly known as Superfund, a Design-Builder should never have to assume any responsibility for pre-existing environmental conditions at a project site. Rather, it should attempt to include contractual language that its sole responsibility will be to immediately notify the Owner if it discovers existing hazardous materials at a project site. Further, a Design-Builder should request that the Owner defend, indemnify, and hold it harmless from and against any claims resulting from the existence and discovery of hazardous materials at a site. Thereafter, the Owner shall be responsible for developing a plan (presumably with its environmental consultants and contractors) for the removal and remediation of the hazardous substances.

An Owner will probably insist upon contractual language that requires a Design-Builder to assume full responsibility for the remediation and removal of hazardous substances that the Design-Builder or its subcontractors introduces to the project. The Owner will insist that the Contractor or subconsultant defend, indemnify, and hold the Owner harmless from and against any claims resulting from hazardous substances that the Design-Builder introduced to a site. However, the Owner may also insist on a similar indemnification for pre-existing contaminants if the Contractor or subconsultant discovers and identifies the substances, fails to timely report its discovery, and subsequently spreads the contamination to further areas.

AGREEMENT BETWEEN THE DESIGN-BUILDER'S CONTRACTOR AND ENGINEER

If a Contractor is leading the Design-Build team, it will enter into an agreement with an engineering firm to prepare the design. The agreement between the Contractor or Design-Builder and its Engineer will define the allocation of risks and rewards between them. Unlike the more familiar and comfortable design-bid-construct positions that have been staked out over the years, the roles and relationships of the Design-Build project participants are open to reinterpretation and definition. Under the Design-Bid-Build project delivery mechanism, the Engineer and Contractor often found themselves in adversarial postures over project issues. Through the Design-Build method, their posture must be as allies, and the arrangement will be most successful if the alignment of the risks and rewards of the two parties are parallel.

This is not to say that tension between the Engineer and the Contractor will not occur. In fact, much has been written about that tension and the Engineer's professional duty to "protect the public" conflicting with the Contractor's objective of maximizing profits. Not only is this tension a cultural shift for all the parties involved, but it is also a shift in legal responsibility. In contrast to the Design-Bid-Build method, the Engineer in a Design-Build contract owes its duty of loyalty to the Contractor, not the Owner. In addition to the realignment of loyalty required by the Design-Bid-Build project delivery mechanism, the Engineer may now be required to perform its services with a heightened awareness of cost, schedule, and profit.

The risk and reward structure between the Engineer and the Contractor on a Contractor-led Design-Build arrangement can take many forms. These arrangements can range from a simple cost plus fixed fee reimbursement for services provided to a fully speculative equity partner with the Contractor. There is also a growing trend toward Engineer-led Design-Build arrangements. These teaming arrangements are discussed more thoroughly in Chapter 4.

In most cases, the Engineer–Contractor agreement will present a phased approach to the project. The phases correlate sequentially with the project's evolution from preparation of the proposal and securing the project to the design and construction phases. During each of these phases, the agreement must clearly define the scope of services and division of responsibility between the Engineer and Contractor. For a subsurface project, the Engineer's involvement in reviewing the Owner's preliminary design and geotechnical data during the proposal stage, and in lending its expertise to critical decisions on excavation techniques and design requirements are crucial to being selected. A bad decision at this point in the project will carry through into construction, and the risk of which party bears the costs of that decision should be clear.

If, during the proposal, selection, and negotiation stage of a project, the Engineer and Contractor have each agreed to bear its own cost, the Engineer may very well be risking significantly more, relative to its ultimate rewards, than the Contractor. This is because the selection and negotiation process requires a concentrated engineering effort. Even if the Contractor is contributing equivalent resources to the proposal/negotiation effort, the Engineer is still risking more, because the Contractor stands to reap significantly more if the venture is successful. One way to balance this apparent inequity in the risk-reward calculus is for the Contractor to agree to make a success fee award to its Engineer in the event of a successful negotiation and execution of a contract with the Owner.

During the design phase of a subsurface project, additional geotechnical data will be obtained in order to verify assumptions and complete the designs. The confidence an Engineer has in its characterization of the ground is generally directly proportional to the amount of data available. The collection of this data (as described in Chapter 8) can be expensive. Thus the agreement between the Contractor and its Engineer should describe the scope of the investigation to be undertaken and provisions for amending that scope. Both the Engineer and the Contractor need to be comfortable with the decision that the investigation is complete and confidence in site characterization is sufficient. Ultimately, the risk of encountering less favorable conditions than anticipated will probably run to the Contractor. Accordingly, the Contractor should make the final determination when enough data has been collected, and the agreement should provide for additional compensation to the Engineer for performing that investigation.

The Engineer–Contractor agreement should indicate which party (the Engineer or the Contractor) owns and/or can use design materials prepared in connection with the project. It is not unusual for the Engineer to maintain ownership of design materials while the Contractor simultaneously has the unfettered use of the materials in connection with the project. If the Contractor is accorded rights to use the design materials, then it is appropriate for it to agree to indemnify the Engineer for any liability that may extend to it as a result of the Contractor's use for any purpose other than specifically authorized in the Engineer–Contractor agreement. This is especially true in situations in which the Contractor has used the design materials after terminating the Engineer prior to substantial completion or when the Contractor modifies the design materials without the knowledge or approval of the Engineer.

AGREEMENT BETWEEN DESIGN-BUILDER AND ITS SUBCONTRACTORS AND SUBCONSULTANTS

The agreements between the Design-Builder and its subconsultants and subcontractors need not be significantly different than those for Design-Bid-Build projects. Clarity of scope, schedule requirements, and unambiguous quality expectations are all key to these agreements. The major subcontractors and subconsultants should certainly be aware that the Design-Builder is performing on a Design-Build project, and any critical provisions that must flow down to the subcontractors and subconsultants should be clearly flagged. For example, a slurry wall subcontractor may be accustomed to working for a given Owner and allowed to submit a DSC claim through the prime contractor for Owner consideration. If the Owner's agreement with the Design-Builder does not allow these DSC claims, then either that provision should flow down to the subcontractor or the Design-Builder should be prepared to entertain the claim.

Similarly with subconsultants, if the Owner's agreement with the Design-Builder requires a heightened standard of care, and the subconsultant's contribution could impact the design of a critical component, the agreement with the subconsultant should reflect the flow down of that heightened standard. Otherwise, the Design-Builder may be caught with the responsibility for a piece of the design it did not perform.

DETERMINATIVE CONTRACTUAL PROVISIONS

The rights and liabilities between the Owner and the Owner's Engineering Consultant will be resolved in accordance with a handful of determinative contract provisions and the common law of the jurisdiction where the contract takes place. The contract provisions must be drafted with an understanding of the laws of the jurisdiction in which they are intended to operate. A contractual provision seeking to limit liability for certain negligent acts of the Engineer may operate perfectly in a state that favors those provisions but be ignored by the courts as against public policy in a state that disfavors them.

STANDARD OF CARE

Design-Build agreements must set forth the standards of performance to which the Design-Builder will adhere in designing and building the project. This concept is of particular importance for Engineers, because the selection of the applicable standard affects available insurance coverage, among other things. Generally speaking (and in the absence of a contract provision creating a higher standard of care), Engineers are required to provide services that are consistent with those generally provided by other reasonably prudent professionals in the same community. Using this standard, an Engineer has performed negligently (or has departed from the applicable standard of care) when it has done something (or failed to do something) that a reasonably prudent practitioner would do (or would have refrained from doing) under the circumstances. This standard does not demand or expect perfection. It is understood that there is no such thing as a perfect set of plans and that even prudent Engineers will make mistakes. (Please refer to Hatem 1998, Chapter 10, pp. 259–415, for an in-depth consideration of the professional liability and standard of care issues for subsurface projects.)

Engineers should be alert to contracts that impose a stringent standard of care. It is generally not advisable for an Engineer to accept responsibility to perform services "to the highest level of care" or "to the best professional standards." One danger of agreeing to this type of heightened level of care is that any mistake that an Engineer makes—be it an omission from a drawing or an inconsistency between design drawings—becomes a potential basis for a claim for breach of contract. Such a scenario becomes especially problematic when one considers that most professional liability policies provide coverage for traditional negligence (i.e., departure from the traditional standard of care) but not for liabilities assumed by contract.

LIMITATIONS OF LIABILITY CLAUSES

Limitations of liability clauses are an increasingly popular and common way for an Engineer or a Contractor to limit its risks on a project. Unlike an exculpatory clause, which attempts to transfer all the risks from the breaching party to the nonbreaching party, a limitation of liability clause merely limits the magnitude of the damages that can be collected for a party's breach. Typically, these clauses will attempt to limit the liability to either a fixed dollar amount, the amount of available insurance, or to some standard such as the reperformance of the engineering services that gave rise to the breach. The enforceability of these clauses, however, vary widely from jurisdiction to jurisdiction. Some states look favorably upon the clauses while others will hold them invalid. In those jurisdictions where the clauses are favored, several factors affect the enforceability of the clause. First, the clause must be clear and unambiguous as to exactly what liability is being limited. For example, is it the intent of the clause to limit the liability for both breach of contract and tort actions? Second, the limitation must be reasonable with respect to the value of the services provided. Finally, the clause must have been bargained for—that is, the party against whom the limitation of liability clause is being enforced must not have been powerless to negotiate the clause. In the case of most contracts in this book, both parties are sophisticated entities that bargain at "arm's length," and clauses such as this will be found enforceable.

Even in those jurisdictions where limitations of liability clauses are held valid, in certain circumstances, such as gross negligence or fraud, the enforcement of the clause would be considered unconscionable.

Closely coordinated with the limitation of liability clause should be a disclaimer of consequential damages provision. This provision insulates the Design-Builder from the Owner's damages associated with loss of revenue, loss of use, or other business and economic loss resulting from the Design-Builder's negligence or breach of warranty. This type of provision, as well as caps on liability, is meant to protect the Design-Builder from financial ruin as a result of uncontrollable and uninsurable losses.

INDEMNIFICATION

Using indemnification clauses is one way for parties to allocate risks and protect themselves in design and construction contracts. Generally speaking, the party having the primary control or exclusive responsibility for an aspect of the project will agree to indemnify other project participants that may encounter vicarious liability or exposure as a result of being involved in that aspect of the project. For example, on traditional Design-Bid-Build projects, it is not unusual for a Contractor to agree to indemnify the Owner and Engineers for claims arising from the Contractor's construction activities (i.e., matters that are exclusively within the Contractor's control). Design-Build contracts should specify which party, if any, is required to indemnify the other and under what circumstances.

INSURANCE

The issue of indemnification is integrally connected with the issue of insurance. In Design-Bid-Build projects, parties obtain insurance to cover risks inherent in the area of the project for which that party has responsibility. In other words, Contractors obtain insurance for their construction work while Engineers obtain professional liability insurance to cover their professional services. For the most part, these types of insurance coverage are mutually exclusive. For example, performance bond sureties are structured to cover only the work itself as distinct from the design of the project. Professional liability insurance covers project services but not the work of the Contractor. As a result, insurance and insurance-related issues are sometimes difficult to incorporate into Design-Build contracts, which blur the distinction between construction and design. Because of the complexity of this insurance issue, it is advisable for Engineers, Contractors, and Owners to obtain qualified insurance advice prior to entering into a Design-Build agreement.

DISPUTE RESOLUTION PROVISIONS

Disputes are almost inevitable, especially on subsurface projects. Accordingly, it is prudent for all the agreements to consider and prepare for dispute resolution in the fairest and most expeditious manner possible. The default option, always available unless bargained away, is litigation in state or federal court. Prior to the advent of the Design-Build method, most standard form construction agreements provided that disputes between Owners, Contractors, and Engineers would generally be submitted to the American Arbitration Association for resolution. More recently, standard form agreements have modified this procedure by incorporating some form of mediation either prior to or instead of arbitration.

The subsurface construction industry has been generally well-served by the dispute resolution process first promulgated by the Underground Technology Research Council, and later by the American Society of Civil Engineers. This well-known process is founded on three cornerstones:

1. Escrow bid documents, wherein the Contractor lays out its assumption and estimated cost for performing the work;

2. A geotechnical report that is adopted as the project's interpretation of geotechnical data and provides a baseline for determining whether a DSC has been encountered; and

3. Dispute review board (DRB), consisting of three members, one selected by the Owner, one by the Contractor, and the third selected by the other two members.

The board meets regularly with the parties to discuss progress, problems, and concerns, and tours the work as it progresses. When a dispute arises that the parties cannot resolve, it is presented to the DRB, which makes a recommendation on how the dispute should be settled, based on its familiarity with the work, contract documents, respective obligations, and the presentation from the parties. The recommendation is generally nonbinding, but usually leads to a negotiated settlement in the dispute.

In addition to indicating the process that parties will use to resolve disputes, Design-Build agreements should also spell out particulars as to how, when, and to whom a party is to give notice of its claim. In this regard, parties should also consider if they wish to incorporate any time or financial restrictions on a party's ability to pursue a claim. For example, parties can choose that all claims arising from a project are deemed to have accrued at the date of substantial completion. Similarly, parties can agree to apply an abbreviated statute of limitations to claims arising by and between them. They can also choose to limit a party's liability for claims arising from the project. (Of course, third parties not privy to the Design-Build contract would not be bound by such a limitation.) It is not unusual for members of the Design-Build team to limit their liability so as not to exceed available insurance coverage.

TERMINATION

Although parties about to engage on a project are generally optimistic and expect that the project will go forward on time, within budget, and without mishap, it is prudent to plan for the worst and to provide for how the parties will end their relationship should the project not work out for the best. The Design-Build agreement should establish under what circumstances a party can terminate the contract and with what repercussions, and should determine, for example, if the Owner should be able to terminate the agreement without cause and, if so, the Design-Builder's remedies; and if the Design-Builder is entitled to recover out-of-pocket costs, lost profit, termination expenses, or consequential damages.

RECOMMENDATIONS

Recommendation 6-1: Members of the Design-Build team should exchange concerns about proposed terms and conditions prior to finalization of a contract with an Owner and understand the risks posed by the contract, including, but not limited to, schedule and delay risks, onerous terms, and financial penalties, before final pricing occurs and commitments are made.

Recommendation 6-2: Once the project commences, it is critical that the subconsultants and subcontractors to the Design-Builder identify project issues as they occur and provide timely notice to the Design-Builder. In turn, the Design-Builder must take guidance from its team members (when appropriate) and notify the Owner when such issues and concerns are identified.

Recommendation 6-3: The agreements should contain an obvious and unambiguous memoralization of the allocation of risks, which should be clearly assigned through one of the project's contracts. The Owner should understand how the Design-Builder is transferring and mitigating contractual risks through the services to be provided by other members of the Design-Build team.

Recommendation 6-4: Sureties and appropriate insurance should be used to backstop otherwise unacceptable risks. The Owner should also consider establishing contingency accounts to address risks that are beyond the control of project participants. In addition, the members of the Design-Build team offering professional services (i.e., the design engineers) should provide certificates of insurance providing evidence of sufficient coverage for the amount of reasonably foreseeable risk associated with a project.

Recommendation 6-5: Careful early planning of the entire project should be factored into developing the contracting architecture. Of particular importance, in the early phase of the project, is identifying and characterizing the interfaces between contracts. The Design-Builder must translate the entire scope of work negotiated with the Owner into clearly defined segments so that appropriate scope risk is assigned to the proper party and each party clearly understands its role in the entire process.

Recommendation 6-6: The generally accepted form agreements will likely require some modification to fit a subsurface project, and editing and modification should be accomplished by practitioners familiar with the underground construction industry.

Recommendation 6-7: The Owner and the Design-Builder should address (in the contract) an appropriate transfer of risk and responsibility for unique challenges associated with a specific project. To a certain extent, reasonable contingencies can be developed for issues such as material shortages, permitting delays, seasonal interruptions, and accommodations due to public needs.

Recommendation 6-8: For the resolution of disputes, an effective agreement should also include a fair and reasonable process that allows the parties to raise and attempt to resolve their issues and disputes without affecting ongoing project performance and completion.

NOTE: The authors thank Donna M. Hunt, Esq., a registered architect and associate at Donovan Hatem LLP, who assisted in research for the preparation of this chapter.

BIBLIOGRAPHY

American Council of Engineering Companies. 2001. *Multiple Project Delivery Systems: The Design Professional's Handbook: Design-Build Project Delivery.* Washington, DC: American Council of Engineering Companies.

AIA (American Institute of Architects). 1996. AIA Document A191: Standard Form of Agreement Between Owner and Design/Builder. Washington, DC: AIA.

AGC of America (Associated General Contractors of America). 1993a. Standard Form of Design-Build Agreement and General Conditions Between Owner and Contractor (Where the Basis of Payment Is a Lump Sum). AGC Document No. 415. www.agc.org/galleries/contracts. Accessed January 2010.

——. 1993b. Standard Form of Design-Build Agreement and General Conditions Between Owner and Contractor (Cost When the Basis of Payment Is the Actual Plus Fee with a Guaranteed Price Option). AGC Document No. 410. www.agc.org/galleries/contracts. Accessed January 2010.

——. 2007. ConsensusDOCS 400: Preliminary Design-Build Agreement Between Owner and Design-Builder. www.agc.org/galleries/contracts/400%20Redline.pdf. Accessed January 2010.

DBIA (Design-Build Institute of America). 1997. "Design Build Contracting Guide." DBIA Document 510. Washington, DC: DBIA.

——. 1998. Standard Form of Preliminary Agreement Between Owner and Design-Builder. DBIA Document 520. Washington, DC: DBIA.

——. 2009a. Standard Form of Agreement Between Owner and Design-Builder—Lump Sum. DBIA Document 525. Washington, DC: DBIA.

——. 2009b. Standard Form of Agreement Between Owner and Design-Builder—Cost Plus Fee with an Option for a Guaranteed Maximum Price. DBIA Document 530. Washington, DC: DBIA.

——. 2009c. Standard Form of General Conditions of Contract Between Owner and Design-Builder. DBIA Document 535. Washington, DC: DBIA.

EJCDC (Engineers Joint Contract Documents Committee). 1995a. Standard General Conditions of the Contract Between Owner and Design-Builder. EJCDC Document 1910-40. Published jointly by American Council of Engineering Companies, American Society of Civil Engineers, Associated General Contractors of America, and National Society of Professional Engineers.

——. 1995b. Standard Form of Agreement Between Owner and Design-Builder on the Basis of a Stipulated Price. EJCDC Document 1910-40A. Published jointly by American Council of Engineering Companies, American Society of Civil Engineers, Associated General Contractors of America, and National Society of Professional Engineers.

——. 1995c. Standard Form of Agreement Between Owner and Design-Builder on the Basis of Cost Plus. EJCDC Document 1910-40B. Published jointly by American Council of Engineering Companies, American Society of Civil Engineers, Associated General Contractors of America, and National Society of Professional Engineers.

——. 1995d. Standard Form of Agreement Between Owner and Owner's Consultant for Professional Services on Design-Build Projects. Published jointly by American Council of Engineering Companies, American Society of Civil Engineers, Associated General Contractors of America, and National Society of Professional Engineers.

Hatem, D.J. 1998. Professional liability and risk allocation/management considerations for design and construction management professionals involved in subsurface projects. In *Subsurface Conditions: Risk Management for Design and Construction Management Professionals.* Edited by David J. Hatem. New York: Wiley.

CHAPTER 7

Design Development

William W. Edgerton, P.E.
Jacobs Associates, San Francisco, Calif.

Hugh S. Lacy, P.E.
Mueser Rutledge Consulting Engineers, New York, N.Y.

This chapter discusses how design development progresses differently in Design-Build subsurface projects than in Design-Bid-Build projects. These differences arise from the contract roles and obligations of the parties, which affect their behaviors during the design process. For example, a failure on the Owner's part to adapt to the Design-Build delivery method can result in problems such as a fuzzy scope of work or a scope of work so narrowly defined that it leaves a Design-Builder little room for the innovation that is the intended hallmark of the Design-Build delivery method. Potential pitfalls such as these are discussed in detail in the context of the differences between the parties' roles in the two delivery methods. The chapter concludes with recommendations for the parties to adapt their behaviors to the selected delivery method in order to achieve its full benefits.

EVOLVING PROJECT REQUIREMENTS: FUZZY SCOPE

In a Design-Bid-Build contract, the Engineer continuously adjusts its design to the Owner's evolving requirements, which are determined by factors that are investigated simultaneously with the Engineer's design progress. These factors and associated requirements are shown in Table 7.1.

As these requirements are developed (or discovered), the Engineer incorporates the changing criteria into the design documents while respecting the stipulations of the codes and industry standards as they apply to the work. The Engineer's responsibility to incorporate these changing criteria is generally unchallenged by any conflicting duty. The Engineer's duty to its client, the Owner, its responsibility to adhere to the Engineers' Code of Ethics, and its duty to public safety arising from the engineering license are generally all in harmony.

In a Design-Build contract, the Engineer works for the Design-Build entity, which is driven to meet the Owner's requirements in a manner that provides the cost and schedule savings that make this delivery method profitable. The parties do not have the luxury of being as flexible in incorporating changing requirements into the design, because lost time often equates to lost money.

As a result, for the Owner of a Design-Build project to avoid disputes with the Design-Builder about changes to the work, the requirements listed in Table 7.1 must be defined to an adequate level before the Design-Build Engineer is engaged. If the Owner fails to adequately investigate and develop the associated requirements, the scope of the project will become fuzzy as the Owner continually passes down new requirements based on its latest investigations. The

TABLE 7.1   Changing requirements during design development process

Requirements	Determined by
Plans/specifications for anticipated subsurface conditions	Geotechnical investigations
Permit conditions	Regulatory authorities
Utility relocations	MOUs* with various utility companies
Architectural finishes	Historic preservation authorities, public participation
Environmental mitigation	Environmental documents
Restrictions on means and methods	Agreements with the local community, interested third parties, and/or risk evaluations
Building protection	Adjacent property owners

* Memorandum of understanding.

Owner's dilemma in the Design-Build model becomes how to encourage innovation and shed contractual responsibility for changing requirements while minimizing scope changes during construction. The normal way of addressing this dilemma is to establish, in detail, the level of scope included in the tender documents.

LEVEL OF SCOPE IN TENDER DOCUMENTS

One of the important issues facing any Owner preparing to advertise for a Design-Build procurement is what degree of detail to include in the initial documents. In a Design-Bid-Build procurement, these design documents are typically prepared to the 100% completion level by the Owner's Engineer, and thus the Contractor's bid price need contain little contingency to cover undefined scope.

On the other hand, in a Design-Build procurement, the final design documents are prepared by the Design-Builder after it has provided a bid for the project and been engaged to design and build it. At the time the bid price is developed, many of the detailed project requirements are not known. Therefore, the bid must include a certain level of contingency to allow for uncertainties. Understanding this concept, and with a goal of achieving the best price with the least risk, the Owner would be well served, from a business perspective, to develop the Design-Build procurement documents to minimize contingencies. How then should the Owner and its Engineering Consultant determine the degree to which the design should be advanced? Three issues should be evaluated (ACEC 2001):

1. Can the scope be adequately defined by a performance specification? To answer this question, we must understand the terms *adequately defined* and *performance specification*. *Adequately defined* is a subjective determination as to whether the scope is well enough established that an independent observer (or adjudicator) could determine if the contract requirement has been met. Sufficient scope definition allows an estimator to determine the expected cost of performing the work. *Performance specification* is a method of describing a portion of the work by means of its intended end use, as opposed to a method of describing how to accomplish the work (typically referred to as a *design specification*). A good example of a performance specification is a tunnel of a specified inside diameter, length, and location, which must not leak more than a certain amount per unit of tunnel length. The performance specification would not dictate the method of excavation, initial support, or final lining type or thickness. If the work can be adequately defined by a performance specification then it does not make sense for the Owner to have it designed in detail by its Engineering Consultant, because to do so would eliminate the incentive to innovate—one of the advantages of Design-Build.

2. Can the scope be quantified for pricing? One of the Owner's interests is to incite sufficient economic competition to generate competitive bid prices. In order to do this, the scope and the conditions under which the work will be done have to be defined to a degree sufficient for the bidders to know what will be required to meet the objective. If the specific work items (or restrictions) are not defined such that competitive bid prices can be obtained, it is in the Owner's best business interest to have its consultant define it further. The best illustration of this concept within the context of the subsurface Design-Build industry is the need for geotechnical investigations. The feasible design approaches and construction means and methods in any subsurface construction are directly dependent upon the subsurface conditions. If the Owner does not believe in the premise that it owns the ground, and if it does virtually no subsurface investigations, then it will suffer some or all of the following consequences:

- Higher bid prices because of the uncertainty of subsurface conditions and, consequently, the Contractor's lack of confidence that the means and methods anticipated in the bid will be successful;

- Higher bid prices resulting from the pre-bid investigations done by bidders (assuming the time and opportunity) to satisfy themselves of the subsurface conditions;

- Fewer qualified and responsible bidders because some Contractors may be unwilling to risk the contingent exposure resulting from little or no subsurface information; and

- Increased likelihood of claims and litigation resulting from disappointed commercial expectations due to encountering unanticipated subsurface conditions.

These adverse consequences are in addition to the cost of actually performing the post-bid geotechnical investigations, because, if the Owner has not yet performed the required investigations, that cost, in addition to the contingency for uncertainty, will be included in the bid prices. This is a good example of an area where the Owner should compare the money spent on pre-tender geotechnical investigations with the savings attained in the prices submitted by the Design-Build bidders. Due to these sound business reasons, it is recommended that the Owner undertake a significant level of pre-tender geotechnical investigations.

3. Can the Owner or third parties live with substitutions or alternate methods to the design suggestions indicated in the procurement documents? One of the key elements in the success of the Design-Build process is the innovation provided by the Design-Build team in devising solutions to the project requirements. As long as the Design-Builder has design options, then the Owner benefits from the economic competition that results from leaving the design options open. However, if the Owner or third parties do not permit any other options or substitutions, the Owner gains no economic benefit and should advance the design to the 100% level to ensure that it gets the exact product that is required. A good example of a design element that an Owner might not allow to be substituted or altered is architectural finishes, which are often negotiated with local community groups or historical societies. In such circumstances, there is limited innovation potential, as the Design-Builder is unable to provide anything other than what has been specifically agreed. To avoid disputes resulting from the community's expectations, the design should be fully detailed by the Owner's Design Consultant and provided to the Design-Build bidders in the procurement document as a done deal.

Considering these circumstances, in order for the Owner to obtain realistic prices, the design should be advanced further in some cases than in others. Several owners have attempted to address this issue by having their Engineering Consultants advance the design to 100% and then issuing a Design-Build contract—thus attempting to identify the scope in detail and at the same time transfer design liability to someone else. The disadvantages of such a solution are that it (1) minimizes the ability for Contractors to provide innovative solutions; (2) uses, upfront, a lot of the time that was expected to be saved by using the Design-Build method; and (3) fails to get the Contractor's buy-in on the design, thus sowing the seeds of future disputes.

However, it is not essential that work that needs to be advanced to or near 100% completion be done by the Owner's Engineering Consultant. In fact, these scope issues can be addressed by the Design-Builder during the bidding period; however, the Owner must recognize certain disadvantages of having this work provided by the bidders.

First, it has to be done by each of the bidders, not just one of them, and this is expensive. And although (in the absence of a stipend paid to the unsuccessful bidders) it can be argued that the Owner only pays for it once because it only hires one of the bidders, the truth is that the economic competition is reduced by the expense of submitting bid prices.

Second, not all the design solutions will be the same. This may not be an issue if the Owner has the flexibility of accepting any of the proposed solutions, but in cases where a certain design has been sold to third-party stakeholders, this could be a factor in bidder selection. If the Owner is restricted, due to procurement regulations or agency policy, from accepting a proposal other than that of the low (price) bidder, then different design solutions may present an insurmountable obstacle. On the other hand, if the Owner has the ability to use a *best-value* procurement model, then evaluation of technical factors can take into account alternative solutions, and the Owner can possibly award to a proposer whose initial price is higher but whose proposed technical components better address other project objectives. This will require a trade-off between technical and price factors.

As in any firm-fixed-price procurement, changes will strain not only the Owner's budget but also the working relationship between the Owner and the Design-Builder. In Design-Build procurement, changes during construction are just as disruptive as they are in the Design-Bid-Build. One might argue they are even more so, because the driving force on the Design-Build project is the schedule. The difference is that, for the Design-Build method, not only is the contract time expanded (to include the design period), but the parties (Owner, Designer, and Contractor) are less familiar with the process. As a result, changes can occur when they are least likely to be expected and most likely to cause disruption.

The bottom line is that the scope has to be developed in enough detail to obtain competitive prices. There are certain advantages to having this done by the Owner or its Engineering Consultant, but it can be performed by the Design-Builder, if the Owner is willing to accept the consequences.

DESIGN-BID-BUILD CONCEPTUAL DESIGN VERSUS DESIGN-BUILD TENDER DESIGN

In the Design-Bid-Build method, before an Engineer begins to develop the conceptual design, the planning phase has typically been ongoing for months while the funding is established, political approvals secured, and so forth. The conceptual drawings are then typically developed without regard to how the project will be built or the construction details, particularly the more complex ones such as the connection between new and old facilities. These details are usually determined later in the design process as the design is advanced to 100%. At each step of the way, the Engineer is compensated fully for the work performed.

In the Design-Build method, however, this entire process is compressed during the tender phase of the project. The Designer on the Design-Build team typically has a month or two to review the tender documents, which for tunnel projects may give no more than an alignment and some information on baseline geotechnical data. During that one- to two-month time period, the Design-Build team must develop a design that is price competitive and meets the design objectives stated—without input from the Owner. Where the objectives are vague, the Designer must use its best professional judgment and develop a rationale for use later if successful. The overwhelming objective in the tender design is to determine the 100% design but without developing the 100% documents. What this typically means is the Designer must put the time into the bid period such that the tender design is good enough for the Contractor to bid from. This usually requires higher-end input from the Designer, since much of the design is developed based upon a feel for what will work and is not verified by calculations and detailed analysis, simply because of time pressures. From this tender design, the price for the work, or bid, can be established. After the project is won, the rest of the design process can proceed by adding design details and clarifying the design for the Owner's approval. The Contractor is likely to resist changes to the concepts set forth in the tender design.

At this stage, it is important for both the Owner and Contractor to recognize that, at bid time, the design is not complete and ready for construction. Sufficient time must be allowed in the project schedule to complete the design process.

FINAL DESIGN LEVEL OF EFFORT

The question sometimes arises as to whether the scope of the final design documents prepared by the Design-Builder need to be as comprehensive as that prepared by the Owner's Engineering Consultant in the Design-Bid-Build process.

In Design-Bid-Build, the permanent project work is designed in detail by a design engineer under an agreement with the Owner. This design is then depicted in the bidding documents, and it forms the basis for bidding (i.e., for the bidders to develop cost proposals). Sometimes, more than one design solution is provided to promote economic competition. In these situations, the drawings might indicate suggested methods and require that details be worked out by the Contractor to meet criteria set forth in the specifications. In other cases, the project specifications may offer the Contractor the use of alternate construction materials and include requirements for each. Both of these methods are effective in enhancing economic competition, since they encourage a certain degree of Contractor innovation and thus permit bidders to be differentiated by more than a productivity estimate.

Another aspect of this approach is that the geotechnical investigations must be all-encompassing to provide enough information so bidders can consider all possible means and methods of construction. For example, when planning a rock tunnel, laboratory data to enable calculation of tunnel boring machine (TBM) penetration rate are typically provided, even though the bidders are allowed the option of excavating using drill-and-blast methods. Temporary works for Design-Bid-Build are typically designed by the Contractor, although in some cases temporary facilities and/or means and methods are designed, specified, and/or restricted in order to minimize the

Owner's risk.* Rather than directing the design of these temporary facilities, the Contractor's role is restricted to constructing the design solution shown in the bidding documents.

In Design-Build, the procurement documents typically provide the concept (sometimes only to a 30% level), design criteria, allowable options, and rules for interfacing with others who need to participate in the design development (other Contractors, permit agencies, adjacent property owners, Owner's Engineering Consultant, etc.). Subsequently, during the final design process, the permanent design is completed, incorporating the specified criteria, industry standards, building codes, and other requirements. Because the Designer works in conjunction with the Contractor, it normally is not necessary for the Designer to develop more than one design solution. Using the Contractor's input to define the means and methods, the final drawings are developed to implement this plan, and geotechnical investigations may be targeted to provide only the information necessary to support the selected design solution. (The cost/risk tradeoffs of providing geotechnical information either before the tender or during the final design have been discussed in the section on "Level of Scope in Tender Documents.") This premise is sometimes cited to support the theory that the Design-Build process does not require the same level of effort for design as does the Design-Bid-Build process.

However, industry experience over the past 10 years indicates that, with certain exceptions, virtually the same level of effort has to be provided in both processes, albeit on different items of work. This is due to several factors:

- The learning curve for the Design-Builder's Designer to become familiar with the design criteria and all the restrictions set forth in various MOUs with utility companies, permitting agencies, and local community groups overlaps with work already performed by the Owner's Engineering Consultant.

- Despite the fact that the Design-Builder only needs one design option, it is frequently necessary for its Designer to provide a number of them, if only for the Design-Builder to base cost comparisons. (See also the discussion in the next section on the Design-Builder's difficulty making design decisions.)

- The specifications can be simpler because they do not need to provide alternate materials in order to encourage competition. Nevertheless, the development of the specifications can sometimes be more time-intensive, particularly when it is necessary for the Designer to include all the quality control and inspection requirements that must be accomplished to ensure that the finished product complies with its design intent.†

---

* This is especially true when other risk allocation methods are used (such as when the Owner is sharing some of the risk through risk allocation provisions, e.g., the differing site conditions clause), and the Owner's interest in restricting means and methods is greater than if the Contractor is taking all the risk. With such risk allocation provisions, the Owner wants some input to these means and methods decisions, because if they do not work, it is at risk for some liability exposure.

† Most traditional Design-Bid-Build specifications do not include such detailed quality control specifications, since the underlying concept is that the Owner's agent (Construction Manager) will inspect and test anything that might be necessary to determine whether the completed work meets the contract requirements. As such, it works against the Owner's interest for the specification to identify the quality control that it will do itself.

- If the Contractor has not charged the Designer with doing the field inspection, the specific verification points must be identified so that others may provide the appropriate inspection.

- The Design-Builder's Designer usually provides detailed engineering for the temporary works, which is not typical on a Design-Bid-Build project.

The only conclusion to be drawn is that the scope of the final design does not vary directly with the selected project delivery method, but is subject to factors that are independent of the delivery method. In some cases, the final design done on a Design-Build project may require as much or more effort than the design done on a Design-Bid-Build project, and vice versa. However, the role that the parties play in design development will differ.

ROLES OF THE PARTICIPANTS

As previously discussed, in the Design-Build process, the Owner's Engineering Consultant does not complete the detailed design but instead provides details to a lesser level of completion. It also prepares a set of design criteria as well as rules for interfacing with other participants. The Designer, which is part of the Design-Build team, completes the permanent design following the rules and procedures established in the contract between the Owner and the Design-Builder. The Design-Builder has a large role in that it must organize all its resources during the design phase, prepare the design and construction schedule, supervise its Designer by setting priorities, provide input to the Designer on cost and constructability, and make decisions on means and methods in a timely manner to facilitate the completion of the design. The role of the Owner's Engineering Consultant during the final design period is to review the details of the permanent design developed by the Design-Builder's Designer to determine that the design meets the Owner's requirements as stated in the design criteria and other contract documents. The details for how this review is accomplished should be spelled out in the contract documents and usually consists of a series of review cycles that permit an evaluation of the progressing design for functionality, durability, and maintenance considerations.

Given these slightly different roles, it is not surprising that there are different perspectives on how these parties should interact:

Owner. Accustomed to participating throughout the design process, the Owner sometimes has a hard time establishing specific criteria upfront and then restricting its role during the Design-Build design process. It takes a much different mentality to establish the rules in advance and then let someone else follow them than to let the rules develop as the design proceeds. As a practical matter, there are sometimes good reasons why design requirements must be developed in a progressive manner. For example, permit agencies need some level of design done in order to establish their own requirements, and community groups may not be able to agree on architectural finishes until the design concept is advanced to a higher level than 30%. If there are a sufficient number of such instances, the Owner has to ask serious questions as to whether the Design-Build approach is well suited for the specific project.

Owner's Engineering Consultant. Because these consultants are designers themselves, it is sometimes difficult for them to review others' design solutions without objecting on the basis that "I wouldn't have done it that way." It is essential that the Owner's Engineering Consultant recognize that its role is not to contribute to the solution (for it is not part of the Design-Build team and thus not knowledgeable of the agreed-upon cost and schedule tradeoffs), but to review designs for compliance with contract requirements—a totally different objective.

Design-Builder's Designer. It must recognize that its client is the Design-Builder, and that its responsibility to the project Owner is by virtue of the Design-Builder's contract, not because of a shared vision of the project design. The Designer's scope is usually fixed, as measured both in time and in cost, and the focus must be on addressing these objectives, rather than the issues arising from the Owner's changing requirements. The other big change for the typical Designer is learning to live with a single-minded focus on schedule, which is the Design-Build team's driving force. Contractors as a group have a schedule discipline that is derived from their pragmatic approach to cost control, which in turn arises from the reality that, for contractors, time is money (Sweeney and Henner 1999). Indeed, one of the benefits of the Design-Build process is the schedule discipline that the Contractor can impart to the entire team, which now includes the Designer. But this schedule discipline is not typically one of the Designer's strongest suits. Thus, one of the more problematic arguments between team members can develop.

Contractor. The Contractor's strengths provide many of the benefits of the Design-Build process. Its schedule discipline, its ability to provide realistic cost estimates early in the design process, and its capability to provide ongoing value engineering are the primary reasons why this process has become so popular. Nevertheless, the Contractor also is faced with challenges that it is ill-equipped to address. The first of these is unfamiliarity with the design process. If the Contractor views design as a production process, like mining a tunnel, and assumes that when the drawings are 50% complete the design is 50% done, then it will be surprised when the design it thought was agreed upon in the first month of work is changed in the last month because of changing requirements or simply because the Designer completed a detailed analysis and cannot "make the numbers work."

The second issue that a Contractor faces is an inability (or unwillingness) to make key design decisions in a timely manner. It has been said that one of the benefits of the Design-Build process is that the Designer only has to design it once, because with ongoing input from the Contractor it knows exactly how the project will be built, and thus there is no need for bidding alternatives. However, if the Contractor postpones the design decision, or asks for an alternate design so that it can develop alternate costs, or has a supervisor weigh in on the matter, this key benefit is lost. On the other hand, many Contractors have design engineers on staff who regularly design temporary excavation support, prepare shop drawings, plan value engineering proposals, and assist during bid preparation. These staff engineers can assist the Contractor's project management staff in adapting to the Design-Build method.

These different roles can contribute to an effective procurement process, especially if the element of collaboration is introduced in the tender period. Using the collaborative process, the Owner and its Engineering Consultant can conduct confidential meetings with each Design-Build proposer, and the resulting design solutions are likely to be more efficient and less expensive than the more traditional approach in which each proposer works in a vacuum to generate a technical and price proposal. The element of confidentiality is important in making this process successful, and this relies upon the development of trust among the participants.

Successful Design-Build projects require that the parties develop some element of trust, which is fostered primarily through communication. One of the best ways to achieve this is a formal partnering program after the contract is signed. In such programs, the goals and objectives of each party are set forth and recognized by the other participants. Appreciation for the interests of the other parties obviously will not change the contract provisions but will help to identify conflicts at an early stage, thus allowing more time for a reasonable solution to be developed.

IMPLICATIONS OF CONTRACT PACKAGING

A multi-disciplinary project can make good use of the Design-Build method because of the various opportunities for innovation in the final design. However, some large projects are broken up into various contract packages intended to make delivery of the entire project more efficient (Edgerton and Davidson 2008). In some cases, this results in *tunnel only* contracts where the scope of work is limited to portals/shafts and the tunnel. In such cases, there is little difference between the Design-Bid-Build and Design-Build methods, since much of the tunnel-only work (e.g., shaft excavation support, portal slope stability, initial tunnel lining systems) is traditionally designed by the Contractor, even on Design-Bid-Build projects. Thus, the level of design produced by the Contractor is similar in both delivery methods. The Owner is advised to take contract packaging considerations into account when determining delivery methods.

DESIGN LIABILITY

While participants in the Design-Build process have different *roles*, Contractors and Engineers also have slightly different *responsibilities* and *liabilities* than in the Design-Bid-Build process. To summarize, Design-Bid-Build liability includes the following:

- The Engineer is not required to produce a perfect set of drawings but only work at a level that is standard in the profession,
- The Contractor is not required to provide an end product that meets the Owner's needs but only to perform to the requirements as stated in the contract documents, and
- The Owner warrants the sufficiency of the plans and specifications for the purpose intended.

This traditional split of responsibility sometimes leaves the Owner uncertain about whether the design will meet its objectives.

In the Design-Build process, the Design-Builder assumes a greater degree of liability because it typically warrants the sufficiency of the design for the purpose intended or at least for the performance-based requirements typically set forth in the Design-Build procurement documents. It is the degree to which the Owner's purpose and need are set forth in the procurement documents that can result in disputes over design liability.

As the final design documents are produced, the Design-Builder's Designer stamps them as the "Engineer of Record." If the completed construction does not meet the Owner's needs due to either function, agreed-upon restrictions during construction, or other reasons, then the Engineer of Record may be held liable for revisions. At least three issues specific to the Design-Build process can affect this potential liability:

1. Unclear design criteria. To what extent has the Owner's "need" been adequately described in the Design-Build tender documents? Or, in other words, do the design criteria established by the Owner enable the design to be developed in detail? On some projects, design criteria do not clearly express and define the project requirements, and restrictions placed upon the construction by third parties are not clearly defined and communicated. In such cases, the Design-Builder cannot be held to what the Owner may have intended if the intent was not made clear in the initial procurement documents.

2. Verification of initial design. To what extent has the Design-Builder been able to verify the initial design as set forth in the tender documents prior to the tender date? It is not unusual for tender documents to provide a suggested design and yet still make verification of the design part of the Design-Builder's scope of work to the extent that it becomes the Design-Builder's own, at least to warrant for sufficiency of purpose. The problem with this approach is that the tender period is relatively short, and the Design-Builder must estimate the contract price by either: (1) believing the suggested design will work and basing the bid price upon it, or (2) going to the expense of verifying the suggested design during the tender period.

Following the first alternative may result in later disputes if it is determined that the tender-based design was not feasible. Although the Owner has presumed that the Design-Builder has taken these issues into account in the bid price, if, in fact, the Design-Builder has not done so, it has nonetheless assumed the risk that the final design can be constructed for the price quoted and in the time allotted. Following the second alternative will result in higher bid prices to cover the cost of the pre-tender engineering work or could result in fewer competitors simply because of the added cost to prepare a tender. This conundrum might explain why Design-Build procurements have been used more on projects where the design is repetitive in nature (vertical construction) than in the subsurface industry, where each project has a unique design solution.

3. Risk allocation. There is a tendency for Engineers working for Owners on Design-Bid-Build projects to prepare designs more conservatively than if these same Engineers were working for Design-Build clients on the same project. It is common for public agencies to request more conservative designs, in part because of their perception of the risk/cost trade-offs. Thus they structure the risk allocation provisions to minimize change orders and are willing to pay more in the bid price simply to avoid the negative publicity associated with costly delays and securing supplemental funding. When these same Engineers prepare designs for private developers or value engineering proposals for Contractors, their clients are willing to accept additional quantifiable risk under certain circumstances, especially when there are substantial potential savings and no increased risks to worker safety. In some cases, this additional risk is quantified and treated as a contingency, and the risk/cost trade-off clearly suggests the riskier approach.

For example, a tunnel was designed to be excavated with a TBM that would reduce the risk of ground settlement and resulting damage to a crossing sewer located immediately above the tunnel. The Contractor proposed a much faster tunneling method that involved higher risk of ground settlement, taking into account the additional cost of constructing a sewer bypass and replacement. In this case, the faster tunneling method was used, and the sewer was damaged due to the ground subsidence and was replaced.

The Design-Build Designer is under pressure to prepare economical designs to satisfy the Design-Build contract requirements, while meeting its own legal and ethical standards as a professional engineer. The Design-Build contract documents may or may not communicate the Owner's level of risk preference, but the Designer must carefully consider this risk and the design warranty when preparing the design solution. This is not a straightforward evaluation, since design and construction risk in the subsurface industry is not always easy to assess, and conditions can vary so abruptly that the degree of risk mitigation using more conservative designs must necessarily be a joint decision between the Contractor and Designer. Therefore, the agreement between the Contractor and Designer must define the risk allocation between the parties and when each party is solely responsible for specific outcomes.

QUALITY CONTROL AND ASSURANCE

It is generally accepted that two elements are associated with quality control: (1) the solution, and (2) minimizing errors and inconsistencies.

Under the Design-Bid-Build project method, the Engineer developing the design for a facility has to answer the key question: "Is the solution optimal for the client?" There are numerous ways of answering this question, but almost all of them require the employment of a team of senior reviewers. These groups are variously called a board of consultants, an independent project peer reviewer, a technical or project review board, or a value engineering panel. Although these entities might focus on different issues, they are all tasked with determining whether a solution is optimal for the client. In making this determination, they frequently must consult the Owner's original stated objectives (and in some cases unstated objectives), examine the level of risk that may be appropriate for the Owner to assume under the circumstances, and apply their industry skills and knowledge of engineering and the construction environment. This review can be done once near the end of the process or throughout the process at specified intervals. In the latter case, the review board can be viewed as an integral part of the design team, instead of as an independent reviewer (Elioff and Edgerton 2004).

For the Designer developing the design as a part of the Design-Build entity, the quality control task is not as daunting. In fact, only one question must be answered: Is this solution the most cost-effective for the specified criteria? The Design-Build Contractor will answer whether it is the most cost effective, leaving it to the review team to determine whether the solution meets the specified criteria. The team of senior reviewers in a Design-Build project may be called the office of the independent reviewer, the proof engineer, or the check engineer. They will evaluate the contract requirements, compare them to the proposed design solution, and reach a conclusion as to whether or not they comply. Note that this process does not require an understanding of the Owner's original intent or a determination of whether the solution is the optimal one for the Owner's needs.

Minimizing errors and inconsistencies in the documents (the second element of quality control) requires a process of checking the documents and cross-checking between different disciplines to ensure consistency that will limit misinterpretations of the design intent. In theory, the work involved is the same, irrespective of whether the delivery method is Design-Bid-Build or Design-Build. In fact, in part because the Design-Build agreement places some measure of design liability upon the Contractor, the Contractor of a Design-Build project is just as interested in quality control as the Owner of a Design-Bid-Build project.

In practice, quality control for design services using the Design-Build delivery method has emphasized process more than results. The Design-Build agreement typically specifies the process by which designs are checked, the qualifications of the checkers, and sometimes even the colors of the pencils used in the review process. This represents an attempt by the Owner and its Engineering Consultant to specify quality control by requiring a detailed process. Nonetheless, a greater degree of efficiency is likely to be obtained if the Design-Build team establishes its own quality assurance plan, perhaps as a part of the technical proposal, as opposed to having the detail specified in the request for proposal (RFP) documents.

EFFECTIVENESS AND EFFICIENCY

This section is not intended to compare the overall effectiveness of the Design-Build process with Design-Bid-Build but to make specific observations on design-related issues. As a result of the inherent strengths and weaknesses of the participants, it is not uncommon for Engineers working on Design-Bid-Build projects to produce costly, difficult-to-construct, or overly conservative concepts. If these designs have not been sufficiently value engineered during the design phase, then the Contractor is more likely to propose alternatives during construction to make the work feasible to build, which often leads to delays.

When the permanent structures are well established in size, location, and materials before the price is developed, as is the case in the Design-Bid-Build process, the Contractor's efficiency is increased, because the temporary works can be designed after the permanent facility's design is complete. This results in the most economical temporary work, because it can be designed to provide only what is necessary to construct the permanent work. Temporary excavation support can be designed after the final structure sizes and construction joints have been determined in order to minimize the size of the excavation, cost of materials, and number of bracing levels.

With the Design-Build process, because the design and construction are concurrent, the accelerated schedule requires that the Contractor begin installation of the temporary facilities before the final design is complete. Because the permanent structural requirements may be unknown, the temporary works may be overdesigned (stronger and/or larger) than necessary simply to allow for all possibilities for the yet-to-be-designed permanent structure. The Design-Build method allows completion of projects more quickly, simply due to the concurrent nature of final design and construction. Completing a project earlier also provides a substantial financial benefit to the Owner, and may offset potential cost penalties associated with overdesigned temporary facilities.

Most Contractors are not familiar with the design risks that the Design-Builder accepts. As is the case for all uncertain elements of work, it is likely that the Owners are paying a premium to Design-Builders for their acceptance of risks that are not well understood. As the Design-Build method becomes more familiar to a greater number of Contractors, this premium may decrease.

As is the case with any price-based procurement, the price paid for the goods (project) will vary indirectly with the number of bidders. (A higher number of bidders generally yields a lower price.) If a Design-Build process limits the number of competitors, it should be expected that the prices will be higher than if competition were not limited.

SPECIAL ASPECTS OF DESIGN-BUILD FOR SUBSURFACE PROJECTS

For those readers familiar with the use of aboveground Design-Build, it may useful to point out how subsurface projects differ.

In the subsurface project, every project has a unique design solution, which, to a large extent, is determined by the ground conditions. Although construction means and methods may differ between Contractors, the primary factor that determines the success or failure is whether the behavior of the ground is the same as anticipated when the price was determined. To a certain extent, the Contractor's risk can be minimized by developing a design solution that is adaptable to different ground conditions, and its cost can be minimized by the use of familiar means and methods. But it should be noted that means and methods (e.g., equipment and labor) that are inherently adaptable to different ground conditions are frequently more expensive to implement than those that work well only in a narrow range of conditions. This is in contrast to aboveground Design-Build projects, in which the design solutions may be similar from project to project. In such cases, the success or failure does not depend upon ground behavior, but on the skill of the Contractor and the Designer in adapting familiar materials and construction to the Owner's design criteria. As might be expected, this differentiation requires that the Designer have a different set of skills than those necessary for producing successful subsurface Design-Build designs.

Much subsurface construction is performed as public works, primarily for public Owners and agencies. These owner agencies typically have strict procurement regulations concerning the negotiation of scope, price, and schedule, which derive from the concepts of equal treatment for all bidders and achieving the best use of public funds. Such public procurement policies make it difficult to evaluate proposals from Contractors whose design solutions are significantly different from the solutions proposed by most others, particularly if the alternative means and methods expose the Owner to a different level of risk. This problem is not as prevalent in aboveground construction, because—even for projects performed for public agencies—the standard risk allocation provisions (e.g., the differing site conditions clause) do not expose the Owner to as much cost and schedule risk as on subsurface projects.

Perhaps because of these differences, the Design-Build delivery method is not as common in the subsurface industry as it is aboveground, particularly in vertical construction (buildings). As a result, the participants are not as familiar with the different processes, and this unfamiliarity affects their behavior.

CONCLUSION AND RECOMMENDATIONS

There are a number of differences between the Design-Bid-Build and the Design-Build delivery methods of subsurface projects: the progress of the design development, the deliverables, and the roles and responsibilities of the parties. These differences result in changes in the parties' behavior which, if recognized, can be addressed during the project, thus enhancing the probability of success.

The following recommendations have been developed to assist Owners and Design-Builders in coping with the issues raised in this chapter.

Recommendation 7-1: In order to minimize bidding contingencies resulting from uncertain scope and to avoid potential liability, the Owner should establish in as detailed a manner as possible the project scope in the tender documents, while still leaving the Contractor with the flexibility to develop innovative design solutions. In particular, the Owner should identify specific details needed to comply with existing third-party agreements and perform sufficient pre-tender geotechnical investigations to enable bidders to evaluate feasible means and methods of construction.

Recommendation 7-2: The Owner should recognize that the Design-Builder's level of effort required to complete the final design is not significantly different from that on a Design-Bid-Build project and should allow sufficient time in the project schedule to complete the design process.

Recommendation 7-3: In order to facilitate communication and identify conflicts at an early stage, the entire Design-Build team should go through a formal partnering program in which the goals and objectives of each party are set forth and recognized by the other participants.

Recommendation 7-4: The Owner is advised to take contract packaging considerations into account when determining delivery methods.

Recommendation 7-5: The Contractor and Designer that form the Design-Build entity must define the risk allocation between themselves and clearly identify when one is solely responsible for specific outcomes.

Recommendation 7-6: To increase the efficiency of the design process, the Owner should refrain from specifying detailed quality assurance processes in the RFP document but should consider allowing the Design-Build team to establish its own quality assurance plan and include it as a part of the technical proposal.

BIBLIOGRAPHY

ACEC (American Consulting Engineers Council). 2001. "Design-Build Project Delivery." Publication #371-00, p. 15.

Edgerton, W.W., and Davidson, G.W. 2008. Getting the right contract package. In *Proceedings of the North American Tunneling Conference.* Littleton, CO: SME, p. 624.

Elioff, A., and Edgerton, W.W. 2004. Design review boards—current state of practice. In *Proceedings of the North American Tunneling Conference.* Netherlands: Balkema Publishers, p. 5.

Loulakis, M.L. "Liability Trends in Design Build," p. 5.

Sweeney, N.J., and Henner, J.P. 1999. The general contractor's perspective. *Design-Build Contracting Claims.* Aspen Law & Business, Fall, p. 127.

CHAPTER 8

Subsurface Explorations

Gary S. Brierley, Ph.D, P.E.
President, Brierley Associates, LLC, Denver, Colo.

Robin B. Dill, P.E.
Program Director, AECOM Water, Wakefield, Mass.

INTRODUCTION

As has been discussed throughout other chapters of this book, one of the most important aspects of any subsurface project is the accumulation and dissemination of subsurface information. On a major subsurface project, this information can impact every aspect of planning, design, and construction, including the protection of adjacent third-party facilities. For a conventional Design-Bid-Build procurement, the Owner retains a team of Engineers, including a geotechnical engineering firm, that is responsible for implementing the subsurface exploration program, for using that information to assist with design, and for incorporating all relevant subsurface information into the contract document. All those Engineers, including the geotechnical engineer, represent the Owner's interest in developing a complete design that will be constructed by a third party Contractor under separate contract with the Owner. In accordance with standard practice, the Contractor must rely solely on the subsurface information provided by the Owner during the bid phase.

Scores of published reports and legal documents have been written about how this process should be implemented and how the subsurface information can and should be used to avoid, minimize, and help resolve disputes over differing site conditions. Because the contract documents for a Design-Bid-Build procurement are based on 100% design, the subsurface database intended to serve the design process serves equally well as a clear basis for the resolution of disputes during construction.

Unfortunately, the rules are less defined for a Design-Build contract. If the Owner elects to use Design-Build procurement for a subsurface project, there are few, if any, established precedents for which party should obtain the subsurface information, how much information should be obtained, and how that information should be made available to prospective Design-Build teams. Some Owners believe the Design-Build team should be responsible for all aspects of subsurface investigation and, in fact, believe that the transference of this risk to the Design-Builder is one of the major advantages of using this method. Although some types of aboveground projects have straightforward subsurface conditions where this approach might be appropriate, such is not the case for the type of project discussed herein.

The authors believe that, for all subsurface projects, the Owner should retain an experienced geotechnical consultant with the necessary skills to represent the Owner's interests in all stages of a Design-Build project. This consultant would work with the Owner in the earliest stages of the project to develop strategies for performing pre-tender subsurface and laboratory investigations; to accumulate and evaluate available data; to present that data and related interpretations in the Design-Build procurement documents; and to oversee construction activities.

Subsurface projects are uniquely influenced by prevailing subsurface conditions, site access limitations, schedule and budget constraints, the project's goals and needs, the potential for third-party impacts, and the large percentage of total project cost represented by the subsurface work. Thus, no cookbook approach exists as to the appropriate level of geotechnical effort during the pre-tender phase of study as performed by the Owner's geotechnical consultant.

This chapter addresses considerations for implementing an appropriate subsurface investigation program for a Design-Build contract, assuming the Owner has retained a geotechnical consultant and that the Design-Builder will retain a separate geotechnical consultant during the procurement phase and subsequently under the Design-Build contract. This chapter also discusses recommendations for the division of responsibilities between the Owner's and the Design-Builder's geotechnical consultants and for the dissemination of subsurface information during the procurement phase of a Design-Build contract. Geotechnical report preparation for Design-Build projects is further explored in Chapter 9.

The discussions and recommendations given in this chapter, as well as in Chapter 9, are intended to apply to tunneling and heavy civil infrastructure projects, where most, if not all of the constructed facility, is belowground. The authors recognize that, for this type of project, subsurface construction represents a large proportion of the total project cost and carries the most risks, therefore warranting a significant investment in geotechnical investigations by the Owner. Nevertheless, many of the concepts presented here can be applied to vertical construction projects and to shallow, open-cut construction using Design-Build procurement.

BACKGROUND

Before it retains any outside parties for a Design-Build project, the Owner must first think about what it wishes to accomplish with the contract document for design and construction. At the beginning of any project, there is no doubt that the project Owner is responsible for all design and construction activities necessary to bring the project to fruition. However, because almost no Owner possesses all the required in-house resources to produce its own facility, a process is begun whereby various design and construction responsibilities are contracted out and whereby selected parties agree to perform services and to accept risk in exchange for remuneration. For a subsurface project, the biggest risk will likely be subsurface conditions and the impact those conditions will have on design and construction.

The Owner of a project wants a finished facility that fulfills all of the needs of the project and provides a public entity or private corporation with long-term reliable service. As such, the Owner wants a facility that has the proper size, shape, and location; adequate structural integrity; an appropriate service life; and low-cost operations and maintenance; and which can be built for the lowest possible cost and within the shortest possible time frame. Interestingly, these requirements for a finished facility are the same whether the project is built above or below ground. For example, if an entire subway system was built aboveground, there would be little need to perform subsurface explorations, except to provide adequate information to design foundations. Now, imagine that same facility, including all stations and other related structures, placed 50 ft below ground surface in the middle of a well-established urban area, and think of the additional reasons why it is necessary to perform subsurface investigations.

First, it is still necessary to obtain sufficient subsurface information to design the permanent facility, including the necessary foundations. Design criteria need to be established for structural loadings, for the impact of water-related problems, and for the use of appropriate construction materials. Geotechnical input for final design is always one of the primary objectives of any sub-surface exploration program.

Second, and equally important, though sometimes overlooked by the Owner, is that none of the finished facility can be built until the subsurface space needed for construction is created, and that this part of construction represents a significant proportion of total project cost and project risk. In order to create this underground space, the ground must be excavated, controlled during the process of excavation, and supported until the permanent facility is constructed. Hence, in addition to final design, all the means and methods of construction, including excavation, ground control, and temporary support to provide the necessary space, must also be evaluated on the basis of information obtained during the subsurface exploration program.

Third, with respect to creating subsurface space, adverse ground behavior encountered during construction potentially impacts a project in two negative ways: (1) the project itself, by adding cost or delaying the schedule; and (2) with possibly greater impact, adjacent and existing third-party facilities. For a typical linear infrastructure project in an urban area, literally hundreds of adjacent utility lines and building foundations may need to be evaluated. In order to conduct this evaluation, the project design team must accumulate additional subsurface information about the facilities themselves, such as the location and condition of all utilities; and the type, size, and depth and condition of all foundations. For many subsurface projects, the need for this type of subsurface information can far exceed that which is needed for project design.

Fourth, depending on ground conditions, the subsurface exploration program must provide information about potentially detrimental environmental conditions and the presence of hazard-ous materials that can impact the work. These impacts can result from natural substances such as methane or hydrogen sulfide, from contaminants such as petroleum product or cleaning solvent, or from stray currents or ground/groundwater characteristics that could result in deterioration of the permanent structures. Any one of these items can have an impact on design, construction operations, excavation disposal requirements, or safety, which can add significantly to the cost for the work. Worse, if these conditions are not discovered until after the work is begun, the associ-ated impact can increase significantly.

Lastly, although it is difficult to discuss within the context of design, safety is part of the rea-son for subsurface explorations. Although in the final analysis the Contractor must be responsible for safety on the basis of what is observed or uncovered during construction, the more ably the Engineer can anticipate those problems the more successful the project. The design of temporary facilities and the provisions for environmental impacts must take into account construction safety insofar as it is possible, based on the results of subsurface explorations.

To summarize, it is necessary to perform subsurface explorations for a subsurface project for the following reasons:

- Assist with designing both the temporary and permanent facilities,
- Assist with choosing construction means and methods,
- Learn about the condition and characteristics of adjacent facilities,
- Evaluate the potential impacts to those adjacent facilities and affected third parties,

- Identify and mitigate impacts from detrimental environmental conditions and the presence of hazardous materials, and

- Provide input on safety decisions.

This important list of considerations for a subsurface exploration program has nothing to do with whether the project delivery method is Design-Bid-Build or Design-Build. These same types of considerations underlie both methods of procurement, and the same questions must be answered and issues resolved irrespective of how the responsibilities for planning, design, and construction shift between the two methods of procurement.

DESIGN-BID-BUILD VERSUS DESIGN-BUILD

For a Design-Bid-Build procurement, the Owner retains an engineering team to perform a 100% design. Thus, all the various aspects of subsurface exploration previously discussed would be incorporated into that design in one way or another. In general, when a Design-Bid-Build procurement is released for bid, the prospective bidders are instructed to rely on the information provided in the contract documents for preparing their bids, including baselines established in the Geotechnical Baseline Report (GBR; see Chapter 9). If subsurface conditions uncovered during construction differ materially from representations made in the contract, then the Contractor would make claims for extra compensation based on the magnitude and impact of those differences. This is the well-established approach that has been used for almost all U.S. subsurface projects that have been constructed using Design-Bid-Build procurement. With this concept in mind, it is now possible to think about what would change in the accumulation and dissemination of subsurface information if the Owner elected to use the Design-Build method for procuring its subsurface project.

First and foremost, *nothing* changes with respect to the required adequacy, performance, and quality of the finished facility. The Owner still wants the right facility at the right location, functioning reliably and dependably for a long time. The Owner's primary long-term goal, therefore, is ensuring a successful finished facility. Hence, sufficient subsurface information must be obtained in order for the Owner to make certain that the final design is acceptable and that construction of the finished facility is performed in a satisfactory manner.

The Owner should also be concerned about the potential for adverse impacts to adjacent property owners. If utilities are broken or adjacent buildings are damaged during construction, the Owner will be confronted with these problems regardless of liability provisions cited in the contract documents. If these problems become serious, then the work schedule will be delayed and costs will escalate, potentially to a significant degree. Although the Owner would like to completely avoid responsibility for investigating and evaluating third-party impacts by passing these responsibilities to the Design-Builder, the authors of this chapter do not recommend this approach for most major subsurface projects in urban areas.

Project Owners must become involved with potential impacts to adjacent third parties and, therefore, incorporate sufficient information into the contract documents to effectively manage these problems during bidding and construction. Clearly, subsurface explorations prior to contract award will be required both to assist the Owner in evaluating the magnitude of these problems and in making sure that the successful Design-Build team has incorporated appropriate solutions into its project approach. The Owner's advisors play a key role in helping to identify potential third-party impacts and in reviewing how the various Design-Build teams propose to minimize the risks of such occurrences.

Once these two issues are addressed by the Owner, however, its ability to have more flexibility with respect to the scope of the geotechnical program becomes apparent for a Design-Build procurement. Most of the opportunities for a Design-Builder to save money and shorten the schedule on a subsurface project rest with the design of temporary facilities and the use of cost-effective means and methods of construction. Design-Builders can be quite innovative in how they schedule the work and control the ground during construction. One of the many advantages of this process is that the Design-Builder can work hand-in-hand with its Designer and geotechnical engineer during this aspect of the project. Once the Design-Builder decides how to proceed, the design team performs only those computations and produces only those drawings needed for that particular approach. Hence, the Designer is able to focus on the selected method and does not have to deal with many potential alternatives.

SUBSURFACE INVESTIGATIONS FOR DESIGN-BUILD

Subsurface exploration programs for complex subsurface projects always take place in phases, and the question for an Owner is which party becomes responsible for each phase of the investigation and what is the proper sequencing to assure a successful project for all parties. In general, the subsurface exploration program would consist of five distinct phases of work:

1. Accumulation of available information,

2. Interpretation and site reconnaissance,

3. Design phase subsurface investigation,

4. Procurement phase subsurface investigation, and

5. Subsurface investigation by Design-Builder.

The Owner should be fully responsible for the first three phases and have partial involvement in the last two phases, while the selected Design-Build team would typically have involvement in the fourth phase and full responsibility for the last. The remainder of this chapter presents a discussion of how the subsurface investigation program for a major Design-Build subsurface project would be implemented. To better understand the responsibilities of the various parties, it is helpful to review the typical timeline and sequence of events as presented in Figure 8.1.

Phase 1: Accumulation of Available Information

Previous Test Borings. This extremely important part of the program consists of finding the logs of previous test borings at the site and information about other projects that were constructed nearby. No matter how or when previous test borings were drilled, it is necessary to make a good-faith effort to obtain that information and to evaluate it relative to the proposed construction project. Very often existing test borings were not drilled deep enough for the proposed project or were drilled in ways that are not appropriate for evaluating complex subsurface projects. For instance, the top of rock for most foundation investigations is often determined by *refusal* of the drill rods, but such information is of little use if a major shaft needs to be sunk into the rock mass. Or it would be unusual during a typical foundation investigation to actually measure the in-situ permeability of soil or rock materials at a site in contrast to a major subsurface investigation, where this type of measurement is common and the output crucial to evaluating potential design and construction problems.

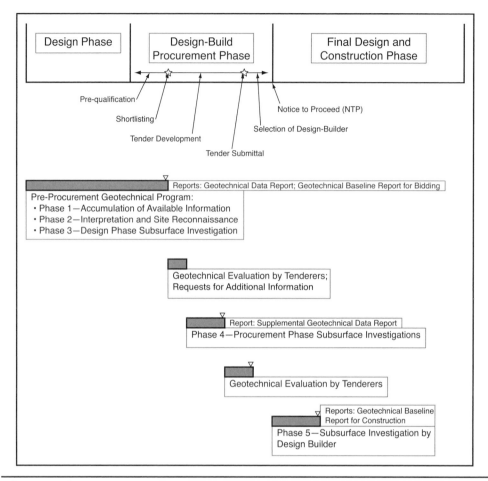

FIGURE 8.1   Subsurface investigation timeline

Previous Construction Activities. Even more important is the investigation of previous construction activities at or near the proposed site. Even anecdotal information about possible construction problems can be useful in evaluating similar possible problems for the project at hand. Discussions with representatives from Owners, Engineers, and Contractors in the area can reveal many aspects of ground behavior and ground response to construction that provide important input about how the subsurface program should be organized and how that information should be accumulated for a major project. Even old newspaper articles provide important information about potential problems.

Existing Utilities. Another aspect of required available information concerns existing utilities. All information about the size, type, depth, and location of each utility should be assembled through contacts with various utility owners. Equally important is information about the condition of those utilities, such as age, type of conduit, and susceptibility to damage. Some utilities are so old and in such poor condition that almost any construction activity would result in potentially serious consequences not only for the project itself but for adjacent property owners. The sooner the Owner knows about the utility situation, the sooner that information can be incorporated into the layout and preliminary design of the proposed facility by the Owner's Engineer.

Building Foundations. Building foundation information and the character of adjacent structures must also be determined. Frequently, the most troublesome are the two- or three-story brick buildings with one level of basement, because such buildings are usually supported on spread footings and can be highly susceptible to damage resulting from differential settlement. In comparison, large high-rise buildings are usually supported on deep foundations, and wood-frame structures can withstand some differential movement. In any case, all foundation information about adjacent properties must be accumulated and input into the planning process as soon as possible after beginning the design.

Historical Structures. In urban areas, a great deal of effort is also expended on deciding what to do about historical structures. In one case, these structures became so difficult to deal with that they were moved rather than underpinned. A comprehensive plan for dealing with all adjacent utilities and existing structures must be incorporated into the Owner's preliminary design and can have important impacts on the scope and magnitude of required subsurface investigations during both the preliminary and final design phases of subsurface investigation.

Phase 2: Interpretation and Site Reconnaissance

Geological Information. At the same time that available test boring information and construction records are being accumulated, the Owner and its geotechnical engineer also must undertake a major effort to develop a comprehensive knowledge of local geological information and history of past site uses, through both literature research and preliminary site reconnaissance. All geologic reports and technical papers should be evaluated for information about subsurface soil, rock, and groundwater conditions at the site. The history of soil sediment deposition, such as glacial, lacustrine, or alluvial, can have a major impact on how the test boring logs are interpreted and how the soil profiles are shown for design and construction purposes. Geologic history is particularly helpful with respect to the location and nature of rock formations, and often contains information about the yield of wells installed in the area for water supply purposes. This information can help indicate both possible water problems and impact to the wells. Geological information can also be the first indication of potentially negative impacts associated with natural phenomena, such as high in-situ stress in the rock, combustible gases, or hydrogen sulfide.

Past Site Uses. Research on past site uses, including parcels adjacent to the site, should also identify scenarios where hazardous materials were stored or used. This process will assist the Owner in evaluating the potential for subsurface contamination and areas that might warrant further study. Files from environmental agencies should be researched for evidence of former releases of hazardous materials or sites undergoing remediation.

Field Reconnaissance. After office research is complete, the Owner's geotechnical engineer should perform field reconnaissance. An important first step in characterizing the site is surficial and structural mapping of available soil and rock outcrops and the location of adjacent water bodies that could be potential recharge sources for groundwater inflow. Bedrock lithology, as well as information about the bedding and jointing characteristics of various rock formations, should be documented during this mapping effort. For major geologic features such as faults, karstic structure, or geologic unconformities, geologic mapping can be one of the best sources of information and the only way whereby specific test borings can penetrate these features as part of the subsurface investigation program. In some cases, the results of geological mapping operations have led to major changes in how the project is configured in an attempt to avoid potentially troublesome ground conditions. Some rock outcrops have been mapped as far as a mile or more from the actual site in an attempt to describe rock characteristics that are difficult to evaluate from test boring results alone.

Phase 3: Design Phase Subsurface Investigations

Once the Owner's geotechnical engineer completes the first two phases of work, it should sit down with the Owner and its design team and have an open discussion about the results found to date, the subsurface-related project risks, and the amount of investment that should be made in a subsurface exploration program. These discussions should include what is needed for project planning, layout, and preliminary design, and what information would be useful for prospective Design-Builders to enable them to develop their approaches and associated costs for the work, including means and methods. In general, there are four primary objectives for the design phase subsurface exploration program.

Investigate the Entire Site. Although the spacing of the investigations would be large in the first stage of exploration, it is important to use the exploration program to investigate the entire volume of soil or rock in which the project will be constructed. Sometimes, for instance, it is advisable to drill the design-phase test borings much deeper than thought necessary, both to investigate ground conditions below the project and to investigate the possibility of modifying the project alignment to a more feasible layout. Other times, it may be advisable to drill test borings outside the project alignment to check for geologic conditions or to investigate possible sources of environmental contamination, as well as third-party impact issues. It is during this design phase program that *all* potential geotechnical problems should be discovered for evaluation during design and for scoping additional subsurface investigations during the later stages of investigation.

Perform In-Situ Tests. Extensive in-situ testing should be performed in the design phase borings for the characterization of important ground properties and engineering parameters. Such testing might include pressure meter testing, downhole or crosshole geophysics, or in-situ soil or rock permeability testing. Also, large-volume geophysical investigations should be employed at this time in order to investigate the largest possible volume of soil or rock for the least amount of money. The in-situ testings should be designed to obtain sufficient information to characterize likely ranges of various ground properties and parameters, and should be commensurate with the degree of associated construction risk.

Perform Groundwater Studies. Depending on the importance of groundwater issues to the project, almost *every* test boring drilled during the design phase exploration program should be used to investigate the condition of the groundwater regime. Multiple position piezometers should be installed in some borings to check for confined aquifers or to facilitate pumping tests. Little can be learned about seepage properties of soil or rock materials merely by studying the static groundwater conditions. If major impacts on the groundwater regime are possible during construction, or if major dewatering efforts are anticipated, then pumping tests should be used during this design phase in order to investigate groundwater conditions. Pumping tests show which soil and rock layers are hydraulically continuous and which layers would be difficult to dewater. Observation wells can also be used to check for groundwater responses following rainstorms, during seasonal variation, and in tidal areas. At one site, significant daily variations in the water levels for one set of design-phase test borings indicated a tidal impact that led to discovery of a major geologic fault and the possibility of saltwater infiltration into a tunnel during construction. If ground freezing techniques are a possibility, then subsurface explorations should focus on measuring groundwater velocities, in addition to the more well-known parameters, including hydraulic conductivity, transmissivity, and storativity.

Perform Laboratory Testing. As with in-situ tests, a significant program of laboratory testing should also be performed during the design phase subsurface exploration program. All work associated with soil and rock classification testing and the determination of important soil and rock engineering properties should be undertaken at this time. In addition, all environmental laboratory tests needed to evaluate possible negative environmental impacts to construction, to adjacent third parties, for safety, and for disposal of excavated soil and rock should be performed; and material properties for spoil materials and groundwater pumped from the site should be established. Most laboratory testing should be performed early in the program, both to discover potential problems and to make certain that the test results are available in time for publication of the Geotechnical Data Report (GDR).

In summary, the design phase subsurface exploration program is a vital part of the ground investigations for a major subsurface project and should be completed by the Owner at the beginning stages of design and well in advance of procuring a Design-Builder. With this program, explorations should be used that penetrate the entire volume of ground proposed for construction, that investigate important in-situ soil and rock properties, that thoroughly investigate the in-situ condition for the groundwater regime, and which provide the beginning of a reliable compendium of laboratory soil and rock test results for classification purposes and for determination of important engineering properties. Upon completion of this phase of the work, realizing that the accumulation of available information and geological studies are also ongoing, the project Owner should have sufficient information to identify all major problems that need to be evaluated for the project, to begin design work for the finished facility, and to establish the scope and magnitude for the subsequent stages of subsurface investigation.

Phase 4: Procurement Phase Subsurface Investigations

After the Owner has short-listed the Design-Build teams based on qualifications and provided them with tender documents, including the GDR and the GDR for bidding, the first order of business should be for the Design-Build teams' geotechnical consultants to evaluate the geotechnical information database obtained thus far on behalf of the Owner. Once this evaluation has taken place, the Owner should meet with each Design-Build team and ask two questions about the adequacy of the subsurface exploration program:

1. How do you feel about the work that has been done to date, and what additional investigations do you believe are required in order to finalize your tender submittal and cost proposal for the work?

2. What scope and magnitude of subsurface investigation do you intend to incorporate specifically into your proposal for the work for your final design and means and methods, should you be successful in being awarded the contract?

For instance, assume for the moment that a particular project involves the construction of a major shaft in an urban area and that the ground conditions include urban fill, medium stiff clay, a layer of sand, and weathered rock. Three Design-Build teams have been short-listed and asked to propose on the project, and each team has vast experience in three different methods of construction: dewatering, freezing, and grouting.

Each method could be made to work at the site and would involve a myriad of decisions about ground conditions and third-party impacts. If the site is surrounded by sensitive structures, then dewatering could be a problem if it resulted in consolidation of the clay, in which case freezing or grouting might be a better alternative. Interestingly, the subsurface exploration program needed for the investigation of a freezing or a grouting program is quite different in that freezing is largely independent of soil stratigraphy details, whereas numerous detailed laboratory tests might be required to evaluate the groutability of both the sand and the weathered rock deposits. Each of these issues then would be the object of discussions between the Owner's team and the various Design-Build teams, and each method would lead to various conclusions about the finalization of subsurface explorations for the site. For a major project, several sets of these types of discussions might be required for different portions of the project. The Owner's team would then carry out the agreed upon procurement phase subsurface investigations so that each tenderer had sufficient information to assess the efficacy of its preferred method.

Also of significance at this point in the project, however, are all the decisions that need to be made about third-party impacts, possible adverse ground reactions, the provision of temporary support, the evaluation of appropriate means and methods, potential environmental impacts, and other issues associated with safe construction practices. Hence, the procurement phase subsurface exploration programs are needed to obtain sufficient information to validate conclusions drawn from the design phase program and to provide detailed subsurface information needed to prepare subsurface profiles and to resolve specific questions raised as a result of the work to date. For instance,

- Is it better to underpin a historic structure or to modify the project layout?
- Is it necessary to perform additional detailed pumping tests to learn more about the aquifer properties in preparation for a major dewatering program?
- How many additional laboratory tests are needed to evaluate the engineering properties of a weak clay shale that shows signs of deterioration during construction?
- Now that the final location of the main access shaft and intermediate production shafts have been established, how many additional test borings are needed to provide information at these sites?
- How many borings and laboratory tests are needed to investigate the sites of potential environmental contamination?

In addition to satisfying the needs of project design, it is also important that the procurement phase subsurface investigation provide a reasonable amount of geotechnical information for prospective Design-Builders to enable them to prepare cost estimates for their intended means and methods, as well as to allow them to assess the risks associated with potential third-party impacts. Hence, the procurement phase exploration program should be designed to narrow the degree of subsurface uncertainty for the shortlisted Design-Builders. For this reason, this phase of subsurface explorations must be performed prior to the tender development, as shown in Figure 8.1. Based on specific project needs as discussed with the shortlisted Design-Builders, the Owner's geotechnical consultant would then complete the procurement phase of subsurface explorations, including additional test borings, in-situ testing, groundwater studies, and laboratory testing, and publish the Owner's Supplemental GDR, as shown in Figure 8.1. The Owner's Engineer would also use the procurement phase subsurface information for the completion of project design, if necessary.

Phase 5: Subsurface Investigation by Design-Builder

The Design-Builder selected for the project is responsible for absorbing and understanding all the information obtained by the Owner and for performing a final design phase subsurface exploration program specifically tailored to the Design-Builder's proposed means and methods. The scope and magnitude of its program would have been established by the Design-Builder during the procurement process, with appropriate costs incorporated in its price for the work. Ideally, the majority of the geotechnical program would have been completed during the preliminary design and procurement phases of the project, such that the Design-Builder has a good understanding of subsurface conditions and project risks and to establish reasonable but not excessive contingencies in its pricing for the work. However, the scope of the final round of subsurface investigations by the selected Design-Builder can vary substantially, depending on selected means and methods, third-party issues, and the inherent variability of ground conditions. It is certainly conceivable, particularly on smaller projects or where the preliminary design and procurement phase subsurface program has answered most of the Design-Builders' questions, that no additional investigation is necessary.

It is important to note that the Owner is in no way off the hook at the time of contract award with respect to subsurface responsibility, as an additional round of subsurface investigation remains to be completed, this time by the Design-Builder's geotechnical consultant. Hence, in order for the Owner to close the loop on this important aspect of the work, it must accomplish two additional objectives:

1. The Owner's Engineering Consultants must become involved with and oversee accomplishment of the Design-Builder's subsurface exploration program to make certain that it is accomplished in a proper manner and in accordance with the contract requirements; and

2. The Owner must review and, at least, comment on reports prepared by the Design-Builder resulting from that work.

Similar to the contract negotiation, this process of coordination with the Design-Builder's geotechnical consultant will require efforts from both parties. In general, the Design-Builder would be required to produce a package of factual information about soil, rock, and groundwater conditions for comparison with the Owner's package. This comparison would include specific discussions about potential impacts to design of the finished and temporary facilities, about the selection of construction means and methods, and, most importantly, about whether the Design-Builder sees any indication at this point to make a claim for differing site conditions.

It must be recognized that a Design-Builder could make a claim for differing site conditions as a result of information uncovered during this phase of the subsurface exploration program. The Design-Builder might discover rock where none was indicated before or uncover environmental contamination or some unanticipated aspect of the groundwater regime. This possibility and the impact that such a claim could have on the project is one of the biggest disadvantages of the Design-Build method of procurement. Clearly, such a claim could affect cost and scheduling and could introduce a highly contentious element into the work at the early stages of mobilization. However, as stated before, at this point in the project, there should have been an adequate amount of geotechnical work performed on the owner's behalf to reduce this risk to a reasonable level.

It is strongly recommended that all the geotechnical data the Design-Builder accumulates be incorporated directly into the contract documents. In essence, the Design-Builder would be contractually obligated to take into account geotechnical data for the site that both the Owner's and the Design-Builder's teams accumulated. In all cases, the Design-Builder should be contractually bound by reliable geotechnical data no matter how that data was developed. As mentioned, it is assumed that the Owner's representatives observed the performance of the Design-Builder's subsurface exploration work and are aware of the usefulness and appropriateness of that information under the contract.

CONCLUSION AND RECOMMENDATIONS

Few project or legal precedents exist for a subsurface investigation program for a major subsurface project using Design-Build. By their very nature, subsurface projects contain large risks for both the Owner and the Design-Builder, and a significant investment in subsurface investigation is warranted. Clearly, it is well within the Owner's sphere of responsibility to implement a program for accomplishing this task, and the authors believe the process discussed in this chapter will result in a successful project outcome for both the Owner as well as the Design-Builder.

The success of this subsurface investigation process depends to a large degree on a number of factors within the control of the Owner. These include

- The Owner must recognize its responsibility to perform a substantial amount of geotechnical investigation in advance of procuring a Design-Builder, and it should budget accordingly, including allowing enough time in the procurement schedule for completion of this process;

- The Owner should retain experienced geotechnical design professionals who have direct experience in Design-Build projects to represent its interests in performing this work and should work collaboratively with them throughout the project; and

- The Owner should involve its geotechnical consultant in an interactive discussion with prospective Design-Builders (and their geotechnical consultants) about the content of the Owner's geotechnical consultant's preliminary design phase subsurface investigation program, the additional needs for subsurface investigations in order to finalize proposals for the work, and the content of the final design phase subsurface investigation that each team proposes. This interactive discussion should be incorporated into the procurement process and will add tremendous value to the project if done in a proper manner.

Clearly, the accomplishment of these tasks requires a lot of work and thought by both the Owner's and the Design-Builder's respective teams of professionals. The success of this approach will also depend on the Owner's commitment to allotting enough time during the procurement process for the necessary communications with prospective Design-Builders and for the Owner's geotechnical consultant to perform its portion of the preliminary design and procurement phase subsurface investigation programs and present the results of these investigations to the Design-Build tenderers prior to final proposals. The Owner must generate sufficient subsurface information to fulfill its responsibilities under the contract and to make certain that the Design-Builder does the same. This process of working together should be highly beneficial to the outcome of the work because it involves input from many different teams of professionals and an extensive discussion of geotechnical problem-solving between the Owner and the winning Design-Build team. If this process of input, discussion, and collaboration is done in a proper manner, then the probability for project success using the Design-Build method of procurement should be high.

The authors' recommendation for best practices regarding subsurface explorations for Design-Build projects is to follow the steps and suggested sequence as presented in Figure 8.1. The recommended steps are briefly summarized as follows:

Recommendation 8-1: *The Owner's geotechnical consultant should perform a phased subsurface investigation program during design. This program should address the needs of both the Owner's Engineer and the Design-Build team.*

Recommendation 8-2: *All geotechnical data obtained should be made available to shortlisted Design-Build teams prior to tender development. Refer to Chapter 9 for detailed recommendations regarding format of reports.*

Recommendation 8-3: *Design-Build teams should retain their own geotechnical consultants during the procurement phase to assess the adequacy of the existing geotechnical database provided by the Owner, given the Design-Builder's planned means and methods. Design-Build teams should be allowed to request additional geotechnical investigations during the procurement phase as necessary to finalize their pricing for the work.*

Recommendation 8-4: *The Owner's geotechnical consultant should implement supplemental subsurface investigations during the procurement phase and should provide that information to all tenderers before finalization of their proposals.*

Recommendation 8-5: *It is expected that the selected Design-Build team will need to perform its own geotechnical investigation after NTP. This program should be subject to the oversight and acceptance of the Owner's geotechnical consultant, and the results of this program should be incorporated into the contract documents.*

CHAPTER 9

Geotechnical Reports

Randall J. Essex, P.E.
Hatch Mott MacDonald, Rockville, Md.

Stuart T. Warren
Hatch Mott MacDonald, Los Angeles, Calif.

INTRODUCTION

This chapter addresses a number of different reports describing site conditions that should be considered for preparation and incorporation in the contract documents or bid documents for Design-Build procurement. (Because a number of topics in this chapter are discussed in other chapters in this book, cross-referencing is made to aid the reader in pursuing related topical discussions.)

In Design-Bid-Build contracting, the Owner and Engineer have an extended period of time to address a range of issues, such as subsurface conditions along the project alignment, the protection or relocation of utilities prior to construction, easements to accommodate rights-of-way through public and private parcels, and the potential for encountering contaminated ground or groundwater during the construction. In Design-Build contracting, the Owner may seek to transfer the responsibility for some or all of these items to the Design-Build team. While further discussion of these points is addressed in Chapter 8, this chapter addresses the types of information that should be investigated and documented in reports before the Design-Build award, who should prepare them, and what the various reports should contain.

A significant portion of this chapter deals with recommendations relating to subsurface investigations for purposes of design and construction planning, and with the differences between the reports prepared for Design-Bid-Build versus Design-Build contracting. Readers are referred to Chapter 8 for discussion about how site investigation, subsurface exploration, laboratory testing, and other site investigation activities are apportioned between the Owner, the Owner's investigations consultant, and the Design-Build team.

GEOTECHNICAL REPORTS

Geotechnical Data Reports

The overriding philosophy of Geotechnical Data Reports (GDRs) is substantially the same for Design-Build as for Design-Bid-Build contracting. It is incumbent upon the Owner to assemble all data and information that has been obtained in the course of site characterization efforts and to disclose this information in an organized fashion. Depending on the size of the project and levels of investigation involved, this information could be contained in one or more volumes.

The information should be organized and presented in a GDR so that Design-Build bidders can readily access and understand the extent of exploration and testing carried out by the Owner. As explained in Chapter 8, Design-Build teams may want or need additional exploration at specific locations or to investigate specific site characteristics in order to support their design approach and construction means and methods. Knowing what has been performed to date is critical to completing these assessments.

The recommendation to have a GDR and to include the GDR as a contract document is made irrespective of whether an interpretive report has also been prepared and included in the contract documents. The reason is that a broad range of issues, tied to the Design-Build team's selected means and methods of construction and temporary support, may need to be addressed by the team—items that will not be known to the Owner or the Owner's Engineering Consultant or advisor. It is therefore imperative that all factual information that has been gathered by the Owner be incorporated into a GDR that is included in the contract documents.

To bypass this important element of disclosure could lead to claims by the Design-Build team on well-established legal bases, such as withheld information, superior knowledge, and other legal doctrines. These and other legal ramifications are discussed in Chapter 3. Please refer to selected references at the end of this chapter to aid in the planning and implementation of site exploration and laboratory testing programs, as well as the preparation and organization of GDRs (National Academy of Sciences 1984; Underground Technology Research Council [UTRC] 1991; Essex 2007).

The GDR should include a discussion of the framework or rationale for the site exploration and laboratory testing program(s). It may be that the plan and profile have been established or approximated based upon preliminary design efforts and that a corridor for characterization has already been developed. Particularly sensitive surface conditions may have precluded access for exploration at certain sites that would otherwise have been investigated. This background discussion will aid the Design-Build team in gaining an understanding of any geologic or man-made issues that (1) constrained the exploration effort, and (2) might result in the database not adequately representing the conditions along the project alignment. Communicating this information to the Design-Build teams could affect their desire for additional information and how they approach the pricing of perceived risks.

For example, a number of water conveyance tunnel projects constructed in the western United States involved project alignments through terrain so environmentally sensitive that the number of permitted drilling sites was constrained due to the environmental impact or disturbance associated with the drilling process. The limited opportunity to explore conditions was directed toward geologic lineaments that were indicative of rock mass defects such as faults or shear zones. As a result, the limited database under-represented the better quality and more massive rock mass conditions along the alignment, which were not able to be drilled. Without an explanation of the exploration constraints, the resulting database would have created an overly adverse portrayal of the conditions along the entire alignment.

As discussed in Chapter 8, bidders in a Design-Build procurement process should be afforded the opportunity to either request or carry out additional exploration at locations critical to their planning. Providing a background understanding of what has been done will help guide bidders in understanding what can and cannot be accomplished through supplementary exploration.

An overview of the regional geologic conditions is another key element for the planning and execution of the site characterization effort, as well as for bidders to better appreciate the types of conditions to be encountered from a geologic perspective. These items are of extreme importance for subsurface construction, whether procured under a Design-Build or Design-Bid-Build framework. The GDR should summarize relevant factual information from the records of nearby, recent construction projects. In some cases, the behavior of key subsurface materials on adjacent projects may be the best indicators of future performance—both favorable and unfavorable.

Factual information should be organized and presented in a manner that clarifies the relevant physical conditions within the various geologic strata and describes heterogeneous conditions that represent potentially adverse conditions. Examples might include sand pockets isolated in a silty or clayey mass, or rock mass defects such as bedding planes, joints, shear zones, and fault zones. Factors that influence the design and construction planning of a subsurface construction project include both the range of conditions and the likelihood of encountering each condition, formation, or feature. These nuances should be discernible from the data as presented in the GDR. If not, there may be insufficient information.

To better assist in the quantification of certain material properties and characteristics, some have suggested bar chart and histogram representations that provide a visual understanding of these important considerations as compared to more traditional but less informative data tabulations. Also, because the Design-Builder's Designer will engage the information in a series of design tasks, making the data available in electronic format may facilitate usage of the data.

The reader is directed to selected references at the end of this chapter to aid in the planning and implementation of site exploration and laboratory testing programs, as well as the preparation and organization of GDRs (National Academy of Sciences 1984; UTRC 1991; Essex 1997). The first reference provides valuable guidelines that apply equally to Design-Build and Design-Bid-Build contracting, regarding how much is enough data. Through the detailed review of more than 80 tunnel construction case histories, the level of investment in preconstruction exploration correlated with the incidence for cost overruns and delays. The evidence presented in the 1984 study is a must-read for those planning an effective exploration program.

Design Basis Reports

Design Basis Reports are often prepared by an Owner's Engineering Consultant to blend industry or Owner-dictated design and operational criteria with interpretations and conclusions from the available geotechnical information. In the Design-Bid-Build framework, these Design Basis Reports are often presented as technical memoranda and serve as bases for advancing the design from preliminary to final stages of completion. In some instances, several construction methods might be technically feasible but present different levels of risk of adverse consequences, such as surface settlement or impacts to overlying and nearby structures or facilities. The Owner may choose to preclude one or more methods of construction because of these concerns. Such information should be discussed fully with the Owner's Engineering Consultant during the preliminary design of the project, well in advance of the Design-Build bid phase. This will better ensure that bidding teams understand the boundaries for planning and pricing the work.

In Design-Bid-Build projects, this type of information is generally documented and refined as the design process proceeds, with the results clearly described in the final plans and specifications. As such, these design memoranda are only provided, if at all, for information. In a Design-Build framework, in which the process may not be as advanced, the Owner should provide as much information as possible regarding preliminary evaluations and associated limitations and rationale.

The Design-Build team will have the responsibility to carry forward the final design of the project and will require the background for design and constructability issues and constraints. For

Design-Build projects, it is recommended that design judgments presented at early stages, thus superseded during subsequent stages of the Owner's design process, be relegated to background information. However, the basis for the reference design shown on the Owner's tender drawings and specifications should be presented in some manner within the contract documents. For example, such information might be presented in a section of the Geotechnical Baseline Report (GBR, discussed in the next section) or within a listing of the Owner's technical requirements. If presented in a document other than the GBR, care should be taken to avoid overlap with GBR discussions.

Because the Design-Build team will have based its final design and construction approach on certain assumptions and evaluations, the Owner should require the Design-Build team to document those evaluations and judgments in either a Design Basis Report or a GBR for Construction (GBR-C). Either document would be submitted with the bid and incorporated into the contract documents. In the course of preparing this information, the Design-Build team should consider all relevant issues carefully and thoroughly. In the event that Contractor or Owner personnel changes occur during the work, the Design Basis Report will serve as an important reference document for subsequent review and consideration.

Geotechnical Interpretive and Baseline Reports

From a contractual standpoint, the preparation and dissemination of a GDR is insufficient subsurface information as a basis for bidding and constructing a subsurface project. An important requirement is the interpretation of the information and the factoring of that interpretation into the design and construction planning for the project. All too often, Owners who have simply given the boring logs to bidders have suffered claims, delays, and cost overruns associated with unanticipated subsurface conditions.

Thus, it is vital, whether within a Design-Build or Design-Bid-Build framework, that the parties incorporate some form of interpretive report that applies judgment to the gathered data and not just presents the data itself.

The preferred format and content of interpretive geotechnical reports for subsurface projects has been the subject of spirited discussion and debate in the United States since the early 1970s, when the first contractual interpretive report was prepared for a Washington, D.C., subway tunnel project. Through the years, different concepts and approaches have evolved, as documented by a number of guideline publications. As new approaches have been identified and implemented, the industry has provided feedback on the effectiveness of each approach. The process is ongoing, and further changes should be expected. The latest position of the "swinging pendulum" suggests that a GBR is the preferred format.

Readers are encouraged to review a number of documents summarized in this chapter's bibliography that chronicle this evolution (National Academy of Sciences 1974; National Academy of Sciences 1978; UTRC 1991; Essex 1997; Essex and Klein 2000; Essex 2007).

GBRs for Design-Bid-Build. Within the Design-Bid-Build framework, the GBR objectives are as follows:

- Interpret the database generated during the project investigations and discuss any known gaps so that a total picture of the anticipated subsurface materials and risks can be understood by all parties;

- Identify the likely behavior of the materials to be encountered, based on the equipment, means, methods, and construction approaches likely to be utilized during the construction;

- Describe the impacts that certain adverse conditions could have on the progress of the work and to nearby structures and facilities;

- Allocate between the Owner and Contractor the financial responsibility of dealing with specific risks, so that both parties clearly understand how those risks will be priced or addressed under the construction contract;

- Describe the basis for the design of the project and how the identified conditions, behaviors, and risks have been addressed; and

- Provide a quantified description of the relevant physical and behavioral conditions anticipated under the contract, so that if the Contractor encounters conditions that it believes are different from those portrayed in the contract, the parties and, if necessary, a neutral third party such as a dispute review board, can answer the question: "Different from what?"

The last bullet item is the most important function of the GBR: to provide a common basis for all bidders to prepare their bids and an objective basis to evaluate an alleged differing site condition. A 2007 guidelines document prepared under the auspices of the UTRC and the American Society of Civil Engineers provides additional information on the contents and suggested approaches for GBR preparation (Essex 2007). Table 9.1 is a checklist of information to be considered for inclusion in a GBR.

TABLE 9.1 Geotechnical Baseline Report checklist

Introduction
- Project name
- Project Owner
- Design team (and design review board, if any)
- Purpose and organization of report
- Contractual precedence relative to the GDR and other contract documents (refer to the general conditions)
- Project constraints and latitudes

Project Description
- Project location
- Project type and purpose
- Summary of key project features (dimensions, lengths, cross sections, shapes, orientations, support types, lining types, required construction sequences)
- Reference to specific specification sections and drawings to avoid repeating information from other contract documents in GBR

Sources of Geologic and Geotechnical Information
- Reference to GDR
- Designated other available geologic and geotechnical reports
- Historical precedence for earlier sources of information

Project Geologic Setting
- Brief overview of geologic and groundwater setting, origin of deposits, with cross-reference to GDR text, maps, and figures
- Brief overview of site exploration and testing programs (avoid unnecessary repetition of GDR text)
- Surface development and topographic and environmental conditions affecting project layout
- Typical surficial exposures and outcrops
- Geologic profile along tunnel alignment(s) showing generalized stratigraphy and rock/soil units, and with stick logs to indicate drill-hole locations, depths, and orientations

Previous Construction Experience (key points only if detailed in GDR)
- Nearby relevant projects
- Relevant features of past projects, with focus on excavation methods, ground behavior, groundwater conditions, and ground support methods
- Summary of problems during construction and how they were overcome (with qualifiers as appropriate)
- Conditions and circumstances in nearby projects that may be misleading and why

Table continued next page

TABLE 9.1   Geotechnical Baseline Report checklist (continued)

Ground Characterization
- Physical characteristics and occurrences of each distinguishable rock/soil unit, including fill, natural soils, and bedrock; and degree of weathering/alteration, including near-surface units for foundations/pipelines
- Groundwater conditions, water table depth, perched water, confined aquifers and hydrostatic pressures, pH, and other key groundwater chemistry details
- Soil/rock and groundwater contamination and disposal requirements
- Laboratory and field test results presented in histogram (or some other suitable) format, grouped according to each pertinent distinguishable rock/soil unit; reference to tabular summaries contained in the GDR
- Ranges and values for baseline purposes; explanations for why the histogram distributions (or other presentations) should be considered representative of the range of properties to be encountered and, if not, why not; and rationale for selecting the baseline values and ranges
- Blow count data, including correlation factors used to adjust blow counts to standard penetration test values, if applicable
- Presence of boulders and other obstructions; baselines for number; frequency (i.e., random or concentrated along geologic contacts); size; and strength
- Bulking/swell factors and soil compaction factors
- Baseline descriptions of the depths/thicknesses or various lengths or percentages of each pertinent distinguishable ground type or stratum to be encountered during excavation; properties of each ground type; and cross-references to information contained in the drawings or specifications
- Values of ground mass permeability, including direct and indirect measurements of permeability values, with reference to tabular summaries contained in the GDR; and basis for any potential occurrence of large localized inflows not indicated by ground mass permeability values
- For tunnel boring machine (TBM) projects, interpretations of rock mass properties that will be relevant to boreability and cutter wear estimates for each of the distinguishable rock types, including test results that might affect their performance (avoid explicit penetration rate estimates or advance rate estimates)

Design Considerations—Tunnels and Shafts
- Description of ground classification system(s) utilized for design purposes, including ground behavior nomenclature
- Criteria and methodologies used for the design of ground support and ground stabilization systems, including ground loadings (or reference the drawings/specifications)
- Criteria and bases for design of final linings (or reference to drawings/specifications)
- Environmental performance considerations such as limitations on settlement and lowering of groundwater levels (or in specifications)
- Manner in which different support requirements have been developed for different ground types and, if required, the protocol to be followed in the field for determination of ground support types for payment; reference to specifications for detailed descriptions of ground support methods/sequences
- Rationale for ground performance instrumentation included in the drawings and specifications

Construction Considerations—Tunnels and Shafts
- Anticipated ground behavior in response to construction operations within each soil/rock unit
- Required sequences of construction (or in drawings/specifications)
- Specific anticipated construction difficulties
- Rationale for requirements contained in the specifications that either constrain means and methods considered by the Contractor or prescribe specific means and methods (e.g., the required use of an earth pressure balance [EPB] machine or slurry shield)
- Rationale for baseline estimates of groundwater inflows encountered during construction, with baselines for sustained inflows at the heading, flush inflows at the heading, and cumulative sustained groundwater inflows to be pumped at the portal or shaft
- Rationale behind ground improvement techniques and groundwater control methods included in the contract
- Potential sources of delay, such as groundwater inflows, shears and faults, boulders, logs, tiebacks, buried utilities, other manmade obstructions, gases, contaminated soils and groundwater, hot water, and hot rock

Source: Adapted from Essex 2007.

GBRs for Design-Build. While the objectives of a GBR in a Design-Bid-Build and Design-Build framework are the same, the preparation process for Design-Build delivery requires some adjustment. As stated in Chapter 8, the Owner should still plan to carry out a significant subsurface exploration program, although each Design-Build team and the eventual Design-Build Contractor may request or obtain additional information. The fundamental difference between the two delivery approaches is that the Design-Build team is responsible for the final design and the selection of equipment means and methods consistent with that design. The Owner controls the site, dictates the general alignment, and carries out substantial portions of the site exploration at the beginning of design development, whereas the Design-Build team is responsible for the final design at the culmination of design development. Given that both parties contribute significantly to the design development process, it is reasonable to consider a process where both parties collaborate in a jointly developed GBR. A multiple-stage process that seeks collaboration and concurrence is presented next.

GBR for Bidding. As an initial step in the process, a document is prepared by the Owner's Engineer that is consistent with the reference design and the amount of subsurface exploration carried out during the preliminary design process. This document, termed a GBR for Bidding (GBR-B), is included with the bid documents sent to all prospective Design-Build teams. The GBR-B focuses on the physical nature of the materials and groundwater conditions likely to be encountered, consistent with the layouts and geometries represented in the preliminary design. In this manner, all bidders are provided with a common picture of the anticipated, relevant physical conditions. The document could also describe the general approaches to design and construction of the project components, including the Owner's desire to preclude certain means and methods resulting from risk avoidance, third-party requirements, scheduling issues, or budgeting issues.

Additional site exploration. In the course of developing the site exploration program and carrying out the preliminary design, the Owner and its Engineer should anticipate that one or more Design-Build teams will request additional subsurface information to that planned by the Owner's Engineer. The earlier in the process that such requests are solicited, the more smoothly and effectively the process will follow. Typical requests might consist of a boring or borings to confirm site conditions at a temporary shaft or more detailed in-situ or laboratory testing in support of a specific ground support approach, excavation method, or nearby structure that needs to be protected. Whatever the reason, the process should anticipate these requests and the additional efforts required to obtain and integrate the requested information. In this manner, any information requested by an individual bidder would be obtained and shared with all bidders through the issued GBR-B.

Physical conditions versus ground behavior. The main focus of the GBR-B is to provide baselines of the relevant physical conditions at the site. Although these baselines would be relevant to any bidder and to any variations in design and construction approach, ground behavior is intimately related to excavation equipment, excavation details, and manner and timing of the ground support installation. Depending on the amount of latitude given the bidders, it would be difficult indeed for an Owner's Engineer to anticipate each bidder's design and planned equipment selection and construction approach. However, if the excavation approach and equipment selection were either a contract requirement or otherwise clearly established in the procurement documents, the GBR-B could address ground behavior aspects as anticipated baseline conditions.

Two examples are presented, one for rock tunneling and one for soft ground tunneling. In a hard rock setting, the use of the drill and blast method might be a viable alternative to the use of a rock TBM. Different rock mass characteristics would influence the ability to advance the heading using the two different excavation methods, as well as the behavior of the resulting excavated openings. Certain rock defects, such as bedding planes, joints, or shears, might have relatively minor impact on the advance rates achieved by drill and blast methods but have a more significant impact on overbreak. This impact would extend to the volume of rock required to be removed from the tunnel and the amount of shotcrete or cast-in-place concrete required during the final lining process. In contrast, those same rock mass defects, depending on their orientations, might have a much more dramatic influence on the advance rate (penetration rate) of a TBM but less influence on overbreak. The types of temporary ground support could also be significantly different for the two excavation methods. Thus, a discussion of important rock mass parameters and characteristics must be tailored to the anticipated method(s) of tunnel excavation.

In a soft ground setting, the efficiency and effectiveness of a cutter-wheel machine might depend on different ground characteristics than those affecting an open-face digger shield. Boulders might represent a problem for one approach but not another. Zones or pockets of unstable ground might influence each approach to a differing degree. If the use of pressurized-face TBMs were required to provide superior face control, different soil and groundwater characteristics would influence the effectiveness of two possible pressurized-face machines: an EPB TBM or slurry TBM. Different soil characteristics would influence the need for various soil conditioners, as well as the design of specific equipment components. Again, different means and methods will influence what are and are not important geotechnical considerations.

In either of the preceding cases, evaluations in the GBR-B would be examined by each Design-Build team and either modified or ratified in a GBR-C prepared by each Design-Build team (discussed next). To facilitate bidder feedback and the Owner's review of different Design-Build team submissions, the GBR-B could be structured to solicit specific information from proposers at designated locations in the report.

GBR for Construction. As a part of its detailed design and construction planning processes, each Design-Build team would interpret the various baselines in the GBR-B, consider those baselines in the planning of its design and construction approaches, and then prepare responses to information requests as presented in the GBR-B. The responses from each Design-Build team would be incorporated in the GBR-B and presented to the Owner as its GBR-C. The GBR-C would disclose all information that each bidder relied upon, in addition to what the Owner provided in the GBR-B, such as additional materials testing conducted by a bidder. In this manner, each bidder's proprietary interests would be maintained, and each bidder's baselines of factual information would be brought forward for incorporation in the contract. From a contractual standpoint, the intent is for the GBR-C to supersede the GBR-B.

Owner Review. The Owner should have the opportunity to review each bidder's GBR-C for concurrence and reasonableness. For example, if, in the Owner's Engineer's opinion, the baseline interpretations and statements offered by a particular Design-Build team were unreasonably optimistic, unrealistic, or otherwise incompatible with statements in the GBR-B, the Owner might want to seek clarification or offer modifications. If a price proposal were submitted at the same time as the GBR-C, and those modifications influenced the cost of the work, the Design-Build team would be entitled to revise its pricing. In the event that an Owner was not permitted to enter into financial negotiations, a two-step process could be implemented. A technical proposal containing the GBR-C would be submitted, and any questions and issues could be resolved within the context of the technical proposal. Then, a price proposal could be submitted that reflected all changes and adjustments to the GBR-C. The latest GBR-C version from the preferred bidder would be incorporated into the Design-Build contract and would serve the same role as a GBR within a Design-Bid-Build framework.

RECENT APPLICATIONS

A number of Design-Build projects have been implemented in recent years with some variation of this approach.

Tren Urbano Subway, San Juan, Puerto Rico

Each bidder was provided with a database of geotechnical information and required to propose supplemental investigations to be carried out during the bid period. No interpretations were submitted within the context of baseline statements or descriptions. Prior to the submittal of final bids, all supplemental information gathered by the Owner during the bidding process was shared with all bidders. Each bidder was required to prepare a Geotechnical Design Summary Report (GDSR) and to submit its GDSR with its bid. The Owner utilized this information during the bid review to gain an understanding of each team's perceptions of the risks on the project. Portions of the winning team's proposal were excerpted and included in the Design-Build contract, which contained a differing site conditions clause. The winning team's GDSR was retained as a part of its escrow bid documentation.

Deep Tunnel Sewerage System, Singapore

In this project, a different approach from the Tren Urbano subway was applied to multiple bid sections on the project. The Owner's Engineering Consultant provided a GDR only, with only modest interpretations of the anticipated subsurface conditions on each contract. As a part of the bid process, each Design-Build team was required to prepare a Geotechnical Interpretive Report (GIR), which addressed specific issues, estimated behaviors, and anticipated parameters influencing tunnel heading advance rate. During the evaluation of the bids, the Owner and its Engineering Consultant critically reviewed the interpretive reports. The contract included a form of differing site conditions clause, and the selected team's GIR was incorporated into each Design-Build contract Compared to industry standards, the number of borings drilled in advance of the bid process was low, and each team was required to price supplementary borings at specific locations. Unfortunately, because timetables for the selection of means and methods preceded the completion of the supplementary borings, this information could not be obtained in time to support certain construction planning and equipment selection decisions.

Sound Transit Program, Seattle, Washington

Sound Transit's early procurement approach built upon experience gained by the Tren Urbano project and engaged the tunneling community in seeking new ideas. The north corridor portion of the project, which was to include almost 8.4 km of twin-bore tunnel, was to have been constructed following a Design-Build format. During the feasibility and preliminary design phases of the project, a phased exploration program was completed. Borings were spaced an average of 100 m apart along the tunnel alignment. In addition, six to nine borings were drilled at each of four transit station sites.

Three Design-Build bidding teams were pre-qualified, one of which withdrew during the formal proposal and cost-estimating phase. At the suggestion of the two remaining teams, an additional four borings were drilled. After the selection of the preferred team, an additional 15 borings were initiated based on discussions with the Design-Build team and based on minor revisions in the alignment. All exploration information was presented in a GDR and interpreted in a Geotechnical Characterization Report. A Tender Geotechnical Baseline Report (TGBR) was prepared, which set out baseline conditions for relevant physical conditions, such as boulder quantities and the nature of soil units to be encountered. All three reports were included in the Design-Build contract. Ground behavior issues were not addressed because they would largely be determined by the Design-Build team's selection of means and methods for tunnel and station excavation and support (Robinson et al. 2001).

The teams were requested to read and factor the contents of the TGBR into their bids and to write a companion document to the Owner's TGBR that documented their selected means, methods, and associated ground behavior assessments. The intent of the process, had it been completed, was for the Owner and Design-Build team to agree upon a joint GBR, which would have been incorporated into the Design-Build contract. Although a preferred team was selected, funding issues prevented a construction contract from being issued. Nevertheless, the approach represents a good model for future Design-Build tunnel projects.

Niagara Tunnel Project, Ontario

This project closely followed the three-step process described in the preceding section. A bid document entitled a GBR-A was prepared by the Owner and issued to the pre-qualified Design-Build teams. The GBR-A solicited input from the Design-Build teams in the form of a GBR-B. In this case, the construction approaches and designs were significantly different among the different teams. Despite these differences, the Owner was able to compare the different GBR-B assumptions and assess the compatibility with the different design and construction approaches. The selected team's GBR-B was discussed, modified, and agreed upon by the parties prior to incorporating it into the construction contract as the GBR-C.

Lake Hodges to Olivenhain Pipeline—Tunnel and Shaft, California

The San Diego County Water Authority utilized a Design-Build approach for this project, which involved the construction of a 1.8-km-long tunnel and 61- to 244-m-deep drop shafts. A GBR-B was prepared by the Owner's Engineer and issued with the bid documents. After issuance of a notice to proceed, as part of its design development, the Contractor was required to prepare a GBR-C incorporating interpretations of any new geotechnical data and providing assessments of ground behavior and ground response for its selected means and methods. The GBR-C was incorporated into the construction contract.

Trans Hudson Express Tunnel Project, New Jersey and New York

New Jersey Transit intends to build a new commuter rail link between northern New Jersey and an enhanced Penn Station in New York City, using a number of Design-Build and Design-Bid-Build contracts. For the first Design-Build contract to be awarded, the Owner had its Engineer author a GBR that was included in the bid package. Throughout the tender process, the Owner solicited questions from the different bidders, and each team made a technical presentation. Throughout the bidder feedback and interview process, the Owner issued seven rounds of changes to the GBR. At the completion of the technical review process, the final version of the GBR was utilized as a basis for a fixed-price tender.

UTILITIES

In addition to ground and groundwater conditions, subsurface construction must cope with other sources of uncertainty. When planning for a subsurface project in highly developed urban and suburban settings, it is highly recommended that the Owner anticipate and investigate the uncertainties inherent in subsurface facilities and utilities.

Historical Perspective

Within a Design-Bid-Build framework, utility searches, field identification, and potholing are typical activities performed by the Owner's Engineer. Drawings are prepared showing the utilities known to exist along the alignment, particularly where surface penetrations will be required, such as construction shafts, permanent access shafts and structures, and cut-and-cover excavations for stations or ancillary structures. In this manner, bidders are clearly warned of the utilities known to exist. If utilities are encountered that were not shown on the drawings, delays can result, and the Contractor may have a reasonable basis for a claim. Although contract language may attempt to allocate all risks for known and unknown utilities to the Contractor, the Contractor will win the dispute more often than not. Thus, from all parties' perspectives, it is important to know the location and characteristics of subsurface utilities in advance of excavation.

Knowledge of existing subsurface utilities is no less important to the parties within a Design-Build framework. That an Owner might see Design-Build delivery as a means of shedding the responsibility of investigating utilities is as fraught with risk as skimping on subsurface exploration. The Design-Build team will fine-tune the locations of its construction shafts and access points based, in part, on where it perceives the least number of critical utilities exist. It will plan construction staging and mobilization around the need for the relocation or protection of critical utilities. The better informed the Contractor is about the existence of utilities in advance, the more effectively it can plan its program. Considering the delays inherent in the securing of permits and approvals to relocate unforeseen utilities, particularly if third-party forces need to be involved, the image of a nickel holding up a $100 bill is apparent. Ideally, the Owner should treat the advanced identification of utilities, both subsurface and overhead, with the same level of care as for Design-Bid-Build procurement.

Alternative Approaches

An Owner can consider a number of different approaches with regard to utilities within a Design-Build framework. As discussed, the process should be treated no differently than for a Design-Bid-Build project.

Another approach would be to explore the Design-Builder's needs for surface access, define zones of potential utility conflicts, and award an independent advance construction contract with the specific requirement of potholing and relocating known utilities so as to provide clear access to selected areas (access shafts, cut-and-cover station excavations, etc.). In this manner, the risks of dealing with uncertainties would be limited to a much smaller utilities contract and would not have the more severe impacts of delaying a much larger heavy civil contract. The Owner would stand to improve its risk position substantially.

The results of this activity could be documented in a Utilities Report that would be incorporated into the follow-on Design-Build contract. The Utilities Report would

- Document the effort the Owner has taken to identify and clear utilities from specified areas, including water, wastewater, telephone, buried electrical, overhead telephone and power, fiber optic, newly relocated utilities, and abandoned conduits;

- Document where less effort has been taken and where the risk of encountering unforeseen utilities is greater;

- Include scale drawings showing the locations of known utilities, including areas where more and less intensive investigations have been focused; and

- Establish baseline statements to allocate the risk of dealing with unforeseen utilities—describing risk mitigation measures incorporated in the contract, such as items for delay time, to be competitively priced by the bidder or assigned by the Owner, in the event that delays are due to utility-related issues.

HAZARDOUS MATERIALS

The treatment of hazardous materials follows closely the philosophies related to other subsurface materials and conditions. A number of Design-Bid-Build tunneling contracts have had to deal with the likelihood of encountering hazardous materials. Typical approaches acknowledge that it is too onerous for a bidder to carry the financial risk of encountering and handling these types of materials and conditions and have included separate means of compensating for these conditions beyond the baseline of financial responsibility. A common approach has been to require the Contractor to develop a health and safety plan as part of the base bid and to carry the costs of a health and safety engineer and personnel training program during the project. If groundwater treatment is anticipated, the procurement of a groundwater treatment plant can also be included in the base bid. However, upon encountering hazardous or contaminated ground or groundwater, costs for investigating, handling, and disposing of hazardous or contaminated materials would be subject to additional terms of compensation, either through predetermined bid quantities and prices or through specified time-and-materials guidelines.

It is suggested that a similar approach be followed for Design-Build procurement. If hazardous or contaminated ground is encountered during site exploration, or suspected due to an occurrence on a nearby previous project, the potential should be specifically addressed in the Owner's GDR prepared in advance of the bid. In much the same manner as discussed in the section on "Geotechnical Interpretive and Baseline Reports" for other subsurface information, relevant information should be provided in the GDR to inform bidders of the anticipated conditions. The Owner's GBR-B should also address the issues. Depending on the anticipated extent of adverse materials or conditions, consideration might be given to developing an Environmental Baseline Report for Bidding (EBR-B), which could provide details to a more advanced degree. Alternatively, these items could be addressed in a dedicated section within the GBR-B.

Preliminary designs completed by the Owner should take these conditions into consideration, so that final design development provides appropriate measures to deal with the conditions if and when they are encountered. The bid documents should either identify those measures the Design-Build teams are required to address or, alternatively, request that the Design-Build teams propose how they should be compensated if such conditions are encountered. As for the broader topic of subsurface conditions, the elements critical to keeping a safe workplace for all personnel on the project and to avoiding delays and unanticipated costs for handling, transporting, and disposing of such materials are a clear definition of anticipated hazardous and contaminated conditions, their attendant risks, and how the Design-Build team is to be compensated for dealing with the conditions.

TABLE 9.2   Documents and information for Design-Build contracts

Document/Information	Prepared by	Content/Contractual Treatment
GDR	Owner	Contract document is subject to additions during the bid phase
GBR-B	Owner	Bid document is subject to review/comment by bidders; superseded by GBR-C
Owner's technical requirements (OTR)	Owner	Bid document describes basis for and rationale behind preliminary design and addresses which design and construction approaches have been precluded; information may be incorporated into GBR-B in lieu of OTR
Utilities Report	Owner	Bid document lists utilities that have either been relocated/protected or require relocation/protection under the contract; illustrated in preliminary drawings and subject to additions and discussion during bid phase
Hazmat/Environmental, possible EBR-B	Owner	Bid document is either separate report, incorporated into GBR-B, or included in contract specifications; subject to review/comment during bid phase
Supplemental GDR	Owner/ Design-Builder	Contract document addresses additional exploration during bid phase
GBR-C	Design-Builder	Document confirms means, methods, and relevant ground behavior issues, ratifies or modifies GBR-B, and becomes contract document following concurrence by Owner

CONTRACTUAL CONSIDERATIONS

The tunneling industry has held healthy debates regarding which documents should be included in the contract documents as opposed to documents provided for information only. The objective here is not to resolve this debate but to heighten parties' awareness of the need for Design-Build documentation that is not typically included or considered under Design-Bid-Build procurement. In a Design-Bid-Build framework, a number of items would typically be resolved by the Owner's Engineer in the course of preliminary and final design, and in the production of detailed construction plans and specifications. However, if the details of final design, as well as the selection of construction approach, are to be addressed by the Design-Build team, key assumptions and limitations should be clearly identified as a part of the bid process.

Table 9.2 contains selected reports and topics with suggestions regarding author and treatment under a Design-Build construction contract. This list should be considered as a guideline and point of departure, and may be extended depending on a project's specific circumstances. If in doubt, Owners should document and disseminate the information for the benefit of the Design-Build teams. In all situations, Owners should consult with experienced legal counsel before making final contractual decisions.

CONCLUSION AND RECOMMENDATIONS

Subsurface construction is a risky business. If carried out properly, it can serve the public's interests by improving our transportation, water, and wastewater infrastructure while preserving and enhancing our quality of life on the ground surface. If not carried out properly, subsurface construction has the potential to adversely impact everyone in the process: the Owner, through costly delays and political strife; the Contractor, through reduced profitability and damage to reputation; the Engineer/Designer, through assertions of errors and omissions; and, most of all, the public at large, because it does not get the improvements constructed within anticipated budgets and schedules.

When engaging in a subsurface project, all parties have a desire to reduce their risks. However, an Owner who sees Design-Build delivery as a means of accomplishing total risk transfer is misinformed. Lessons learned from past projects suggest that Owners and their consultants can best manage their risks on a subsurface project by doing the following:

Recommendation 9-1: *A thorough investigation of the subsurface ground and groundwater conditions should be carried out in advance of the selection of a Design-Build team.*

Recommendation 9-2: *Existing underground and overhead utilities should be addressed, including efforts by the Design-Build team to identify and relocate utilities in advance of construction.*

Recommendation 9-3: *All information learned and obtained in the procurement documents should be properly disclosed.*

Recommendation 9-4: *Design-Build teams should be engaged in a bilateral process during the bid phase that achieves a jointly prepared and agreed-upon GBR-C.*

This will go a long way toward reducing the risks associated with subsurface construction—risks that are essentially the same whether one follows the Design-Bid-Build or Design-Build project delivery method.

BIBLIOGRAPHY

Essex, R.J., ed. 1997. *Geotechnical Baseline Reports for Underground Construction, Guidelines and Practices.* Underground Technology Research Council (UTRC), American Society of Civil Engineers (ASCE).

———. 2007. *Geotechnical Baseline Reports for Construction, Suggested Guidelines.* UTRC, ASCE.

Essex, R.J., and Klein, S.J. 2000. Recent developments in the use of geotechnical baseline reports. North American Tunneling Conference, Boston, June.

National Academy of Sciences. 1974. "Better Contracting for Underground Construction." U.S. National Committee on Tunneling Technology.

———. 1978. "Better Management of Major Underground Construction Projects."

———. 1984. "Geotechnical Site Investigations for Underground Projects."

Robinson, R.A., Kucker, M.S., and Gildner, J.P. 2001. Levels of geotechnical input for Design-Build contracts for tunnel construction. Rapid Excavation and Tunneling Conference, San Diego, June.

UTRC (Underground Technology Research Council). 1989. "Avoiding and Resolving Disputes in Underground Construction." Technical Committee on Contracting Practices, ASCE.

———. 1991. "Avoiding and Resolving Disputes During Construction." Technical Committee on Contracting Practices, ASCE.

CHAPTER 10

Construction Phase Issues

Alastair Biggart, OBE, FEng., FICE
Vice President, Hatch Mott MacDonald, Pleasanton, Calif.

John Hawley, P.E.
Vice President, Hatch Mott MacDonald, Los Angeles, Calif.

John Townsend, C.Eng.
Vice President, Hatch Mott MacDonald, Pleasanton, Calif.

Brian Brenner, P.E.
Senior Professional Associate, Parsons Brinckerhoff, Boston, Mass.

INTRODUCTION

By the time construction begins on a Design-Build project, all the necessary contract agreements, interface agreements, and teaming or partnering agreements must be in place. All Design-Build project participants should also have a clear understanding of their respective roles on the project. When construction commences, the various agreements and the parties' understanding of their respective roles will be tested. This is the point at which things can start to go wrong, especially if any of the agreements have been poorly conceived or are not fully understood.

In order to address the issues that may arise during construction on a Design-Build project, a number of basic contracting assumptions have been made, including that the project is in an urban rather than rural setting:

- The Owner has its own Engineering Consultant(s);
- Design responsibility is clearly defined and communicated;
- The Owner has its own Construction Management Consultant;
- Risk has been shared equitably;
- Design of up to 30% has been carried out at the time of the Design-Build contract(s);
- The Design-Build contract for civil work includes a Geotechnical Baseline Report and a differing site conditions clause;
- A performance specification is included as part of the Design-Build contract(s);
- Interface management between facility and systems contracts is addressed;
- Cash flow is addressed and allows for the early design effort;
- The possibility of additional site investigation is allowed for;

- Equitable procedures are in place for permitting, right-of-way, authority approvals, design approvals, changes, and inter-contract coordination; and
- Contracting practices such as dispute review boards, partnering, value engineering, risk analysis, and constructability review are included.

Unless most of these assumptions have been addressed in a practical and equitable manner, neither the Owner nor Design-Builder should consider a Design-Build contract.

Some people describe the relationship between a Contractor and a Designer in a Design-Build subsurface project as an unnatural marriage, with the Designer as the disadvantaged partner. A procedure to allow some measure of Designer independence must be addressed.

A further important aspect is the Design-Builder's incentive to perform. This means that the contract must contain innovative incentives in order to improve quality and cost, but also incentives to come in ahead of schedule. Incentives may contain both bonuses and penalties linked to performance; for example, with regard to safety or quality, shared savings, and liquidated damages.

In all forms of subsurface construction, it is important that means and methods should be left largely to the Contractor. This should not, however, prevent an Owner from prescribing a particular type of tunneling machine or techniques if it is clear from a risk, safety, and quality point of view that this would benefit the project as a whole.

In subsurface construction, a large element of the risk comes from unknown ground conditions. It is, therefore, assumed that at bid stage the amount of subsurface investigation, carried out in advance, is sufficient for the Design-Builder to be able to evaluate the risks relating to subsurface conditions and to make informed choices as to the type of equipment needed to carry out the project safely and at a high quality.

The essence of this chapter is to address the main issues likely to arise during construction against a background of fair and equitable contract practices. This means fair and equitable to all contracting parties, the aim being to deliver to the Owner a quality project on time and within budget while allowing the Design-Builder to make a reasonable profit.

ROLE OF OWNER AND OWNER'S CONSULTANTS

Program Management

The degree of program and project management support required from the Owner's Consultants depends on the allocation of responsibility under the contract to the Design-Builder and the degree to which the Owner requires sophisticated program management and takes this responsibility in-house. This may also be affected by federal or state regulatory requirements.

Construction Management, Resident Engineering, and Quality Control

Subsurface construction has to be done correctly the first time. The cost, difficulty, and risk of having to correct deficient work and the subsequent effect on schedule are much higher for subsurface work than for other forms of construction, with potential impacts on the Owner that cannot be mitigated by the Design-Builder, regardless of the provisions of the contract. Whatever the form of contract, the Owner cannot escape a measure of responsibility for construction safety, both in terms of financial liability and adverse publicity. For all these reasons, it is in the Owner's interests to play an active part in construction management of a Design-Build contract, in the normally accepted sense of oversight of construction quality, schedule, cost, and safety.

Design-Build contracts still require a Resident Engineer to be the point of contact for day-to-day communications with the Design-Builder, although the various functions may have a different emphasis, and the resident engineering team may be much smaller than for a conventional Design-Bid-Build contract. Resident engineering functions will necessarily include quality control and oversight of field operations, contract administration, project controls, coordination and communications, and interface with the Owner and third parties.

In a Design-Build contract, the Design-Builder typically has primary responsibility for many of these functions, depending on how the responsibilities and risks are allocated in the contract, with the Owner's Consultant concentrating on contract administration and quality assurance.

One of the perceived advantages of the Design-Build method is that the Design-Builder takes more responsibility for factors that are legitimately within its control. Quality control certainly falls within this category. However, there is no reason why a Contractor on a conventional Design-Bid-Build contract cannot equally take responsibility for quality control, and, in fact, this is often tried with varying degrees of success. In the case of Design-Build, depending on the contract provisions, the Design-Builder also has a responsibility for setting the quality control requirements (subject to review and acceptance by the Owner), so the Design-Builder has an added reason for taking primary responsibility for quality control. This does not mean that the Owner should have no site inspection. This is still required to check for mistakes that cannot be corrected, and to have independent evidence when contractual claims are submitted.

Quality Assurance (Audit)

Quality assurance is an essential Owner function that can be performed by the Owner's Construction Management Consultant, by the Owner's program management consultant if one exists, or by the Owner's own quality assurance department. Quality assurance audits should cover all parties' conformance to relevant standards, codes, and procedures. On Design-Build contracts, the Owner's role for quality assurance should be verification of the Design-Builder's own quality assurance—that is, an auditing role. Perhaps the Owner or its consultants would want to dig a little deeper and participate in the Contractor's quality control directly—inspection and verification testing—as part of the resident engineering function previously described. Being too hands-off in quality control was a contributing factor in the tunnel collapse at London's Heathrow Airport in 1994.

Schedule and Payment

Regardless of the extent to which the Design-Builder is made responsible for managing its own work and controlling the schedule, the Owner or the Owner's Engineering Consultant responsible for overall program management should review the initial baseline schedule and monthly updates provided by the Design-Builder to ensure that they correctly reflect the work progress and identify actual or potential delays. In many cases, the Design-Build contract will be one of many construction, installation, and procurement contracts in the overall project, and it is important to maintain an overall master schedule to manage the interfaces between contracts and to bring the project in on time.

Payment provisions for a Design-Build contract may be based on a schedule of values as for a conventional Design-Bid-Build contract but with the proviso that pay items should not depend on details of the final design—for example, linear feet of permanent lining installed, not cubic yards of concrete placed. In this case, the schedule of values would be broken down and allocated to schedule activities for the purposes of progress payments. Monthly measurement can be paid off a resource-loaded and costed schedule, though the schedule would first have to be reviewed and accepted by the Owner as a reasonable allocation of contract values. Whether paid off the schedule or merely based on schedule activities, the Owner's Construction Management Consultant verifies the work performed and on-site materials for each progress payment.

An alternative approach that is applicable to some Design-Build contracts is to base payments on defined milestones, with progress payments based on the percentage completed or spread over the performance period with appropriate adjustments if the approved schedule update shows the milestone is slipping. Depending on the procurement process, the payment schedule can be defined in the bid documents or negotiated prior to award. In this case, payment is based automatically on the approved schedule update.

Engineering Support

If the Owner retains responsibility for the site conditions and for the reference design (it would be unusual for the Owner *not* to retain some responsibility for site conditions and for the reference design that it has issued), the Owner's Engineering Consultant may have a role in revising the reference design either because of shortcomings within the design or differing site conditions. Contractual provisions that change the usual allocation of risk may affect this responsibility, but this is a somewhat separate issue from the decision to use a Design-Build procurement approach. Placing some responsibility for handling small changes arising from site conditions and interpretation of the reference design onto the Design-Builder is normal and recommended.

For bid purposes, the Owner's Engineering Consultant will have produced a reference design, which will typically include drawings defining the requirements for the subsurface facility, including space, and perhaps some mandatory elements of the scheme as required by the Owner or by third parties. There will also be a design manual, provided by the Owner, setting out design criteria and parameters plus specifications of quality requirements and control. The Design-Builder will be required to produce the detailed design, including drawings and technical specifications. This will be reviewed by the Owner's Engineering Consultant, which will check all calculations and review any analyses before giving approval.

Value Engineering and Alternatives

After award of a Design-Build contract, there should be limited opportunity for the Owner to participate in the benefits of value engineering. It is a fundamental requirement for successful application of Design-Build that the Owner's requirements and criteria are fully defined in the reference design, and value engineering reviews should be undertaken before the reference design is finalized. However, the Design-Builder could choose to instigate value engineering reviews of its detailed design as it is developed and, in doing so, may make a beneficial change to the reference design. The Design-Build contract could recognize this possibility by providing for review and sharing the benefits of any Design-Builder-instigated value engineering proposal affecting the reference design with a potential cost saving above a certain threshold value.

In a Design-Build contract, the Design-Builder has considerable freedom to propose alternatives to the reference design shown on the bid documents, so long as the mandatory requirements of the drawings and specifications are followed. In these cases, it is the Design-Builder's responsibility to determine whether such alternatives are cost effective, since they do not require a change to the contract or any adjustment in contract price.

The Design-Builder may also propose an alternative that does not fall within the mandatory requirements, either because it gives a perceived advantage to the Design-Builder or Owner, or to deal with a changed site condition. In these cases, the Owner's Construction Management Consultant reviews the proposal and estimates the cost implications and, with input from the Owner and the Owner's Engineering Consultant, checks that the proposal is acceptable and determines what benefit it provides.

Submittal Review

Submittals are required for a variety of purposes:

- To check the Design-Builder's ability to comply with contract provisions regarding quality control, safety, schedule, and so forth;
- To give the Owner's Engineering Consultant an opportunity to check that the design intent has been properly understood and implemented; and
- To document the finished work in terms of quality of materials, construction, and as-built conditions.

For quality control, safety, schedule, and so forth, the role of the Owner's Construction Management Consultant depends on the allocation of these responsibilities in the agreements between the Owner and its consultants, and in the Design-Build contract. If the quality, safety, and schedule are primarily the Design-Builder's responsibility, the Design-Builder will be required to produce procedures for generating, self-checking, and distributing the documents that are mandatory under the contract (e.g., safety plans or job hazard analyses) or are necessary to perform the work (e.g., working drawings and construction work plans), with certain items submitted to the Owner for record only. However, the Owner's Engineering Consultant should perform quality assurance audits to ensure that the Design-Builder is correctly implementing the approved procedures for these documents and that certain critically important ones, such as the safety plan, may still be subject to Owner review. In many cases, these documents will be required as submittals by public agency third parties, and the Design-Builder should handle them directly, though the public agencies may insist on some involvement by the Owner.

The confirmation of design intent is handled at the overall level by the Owner's Engineering Consultant, which reviews and approves the Design-Builder's detailed design. Confirmation by the Design-Builder's Engineer of Record that working drawings and so forth conform with the Design-Builder's detailed design is still required and should follow a procedure specified in the contract or developed by the Design-Builder, and is subject to audit by the Owner.

Documentation of the finished work in terms of quality of materials, construction, and as-built conditions is necessary regardless of the form of construction procurement. Documentation may also be needed for future modifications, eventual decommissioning, and formal safety certification.

Requests for Information

The Design-Builder's requests for information should be routed through the Owner's Construction Management Consultant and dealt with promptly. Many will require responses from the Owner's Engineering Consultant, and for a large and complex project it would be advantageous to have an appropriately qualified and empowered representative of the Owner's Engineering Consultant based in the field office with the Owner's Construction Management Consultant. In this way, requests for information can be properly understood and responded to in a timely manner, instead of the all-too-familiar scenario where the first formal response fails to address the key question.

One of the advantages of Design-Build is that the volume of requests for information directed to the Owner and to the Owner's Engineering Consultant should be significantly reduced compared with a conventional Design-Bid-Build contract, because items such as clarifications of detailed design intent will be dealt with internally within the Design-Build team.

Defective Work/Warranty Obligations

Normally, an Engineer agrees that its services will be performed and completed in accordance with generally accepted professional standards, practices, and principles applicable to the engagement. This does not mean that every single reinforcement bar will fit exactly as indicated on the drawings or that every dimension will be 100% correct, but it does set a level that must be achieved by the Engineer. It is important for the Design-Builder to recognize this and that requests for information, which may identify small inconsistencies in the design, will become the Design-Builder's responsibility to fix without the ability to claim extra compensation from the Owner. Of course, if significant errors or omissions occur, then the Design-Builder's Designer could find itself with significant liability for construction costs if this exposure is not limited or capped by contract.

Configuration Management/Document Control

In this context, configuration management means the control of documentation to ensure that current drawings and specifications are being used. This includes provisions and procedures for the generation, checking, and approval of documents, and a system whereby the users of documents have the current officially released version. The configuration management system must tie together all generators and users of documents. In the case of a Design-Build contract, the challenge to the Design-Builder is increased because the Design-Builder is responsible for producing and documenting the final design.

Change Management and Control

Many of the changes on conventional Design-Bid-Build contracts arise from resolving design coordination issues or making adjustments to the design to recognize or accommodate site conditions. Design-Build contracts should have fewer changes because resolution of most of these issues falls under the Design-Builder's responsibility. In fact, the contract provisions should be written with this in mind so that minor differing site conditions do not require a change to the contract price. A frequent tactic is to require the Design-Builder to absorb the cost of changes below a certain individual value and up to a certain cumulative value, though there is still an administrative load in documenting changes in case the cumulative value is reached.

In Design-Build, compared with Design-Bid-Build, the Owner has less control of the design process and the management of the work and, therefore, less control of the cost of change work. Significant changes may require the Owner to take back more of this responsibility, negating many of the advantages of Design-Build. There is no room in Design-Build for preference engineering by the Owner and its Engineering Consultant, particularly after contract award.

Responding to Claims

A Design-Build contract is no different when it comes to claims than the Design-Bid-Build contract. Claims should be resolved as early and at as low a level in the hierarchy as possible. The Owner's Consultant's role is normally to review the claim and make a recommendation to the Owner as to its merit, so that a response can be made to the Design-Builder no later than 60 days after the claim is submitted. If the Design-Builder does not accept the Owner's denial of a claim, it should proceed without delay through the claim resolution process established in the contract.

ROLE OF DESIGN-BUILDER'S DESIGN CONSULTANT

Design Coordination and Schedule

Once the Design-Build contract is awarded and the negotiations finalized, the Design-Builder needs to establish and agree with its Designer on a schedule for the completion of the design. To achieve the benefits of Design-Build, this schedule must be construction-driven and, for example, may require completion of building or structural foundation design packages prior to completion of the structure design. Although this may lead to uneconomic design, it ensures that the schedule is maintained. Regular coordination meetings with the various discipline engineers and formal review meetings should also be established. Equally important to both the Design-Builder and its Designer is to establish freeze dates for construction means and methods that may impact the design. These may include finalization of temporary ground support; excavation methods; selection of key equipment, such as tunnel boring machines (TBMs); and the method of erection or installation of key elements. If this is not done, some of the Design-Build benefits are lost, and the team runs the risk of cost and schedule penalties as redesigns are undertaken to accommodate Contractor preferences at a later date.

Technical Specifications

The technical specifications prepared by the Design-Builder's Designer can be significantly reduced from those required on a Design-Bid-Build project, because often the latter specifications are written to cover all potential means and methods and eventualities. It is important that the specifications are tailored to both the design intent and the means and methods that the Design-Builder will use to carry out a particular item of work. These specifications will contain basic design requirements—for example, concrete strength—and will also establish quality control requirements. The ability of the Designer to tailor these specifications to the particular design being constructed and using certain means and methods will allow it to require more or less rigorous quality control depending on the design criticality.

Document Control

It is essential that at an early stage the Design-Builder and its Designer agree upon the various levels or stages at which drawings and specifications will be issued for review. Design-Builders may also require a further stage prior to construction that will allow the ordering of long-lead items. Again, if the full opportunity of Design-Build is to be obtained, long-lead items must be identified early so that portions of the design can be completed at an early stage. A comprehensive document control system ensures that preliminary drawings issued to the Design-Builder—for example, only to allow the ordering of critical long-lead items—are not used for construction. This system must also consider Owner review requirements, because the process can be complicated if the Owner requires design submittals at various stages. Owners that are not familiar with the Design-Build process may be reluctant to approve the design of a package until they can see the complete design; that is, they may be reluctant to approve a construction-critical foundation design package unless they are able to confirm the complete design of a structure.

Design Innovation/Value Engineering

One of the major advantages of Design-Build is that design innovation is encouraged and can allow the Design-Builder to provide the Owner with a facility that achieves its requirements in terms of operation, quality, and design life at a lower total cost than with Design-Bid-Build. For example, on the immersed tube Medway Tunnel in the United Kingdom, a design for niches in the roof to house the ventilation fans allowed a reduction in tunnel cross-section and a significant reduction in dredging volume. This allowed the Design-Builder to offer a significantly lower total cost while fully achieving all the Owner's needs. Once the contract is awarded, opportunities diminish due to schedule restraints. The Owner must also recognize that once the contract is awarded and the mandatory requirements are met, the Design-Builder has the ability to design or redesign elements to allow the most cost-effective construction to the Design-Builder's advantage. Value engineering may also be used if the Design-Builder can provide a design innovation that requires relaxation of mandatory requirements, which could result in a shared cost saving.

Level of Design

The Designer often thinks that the level of detail required by a Contractor for a Design-Build project is less than that normally required by an Owner going to bid. This may be true in the case of specifications, because only the methods and materials that the particular Design-Builder will use need to be specified, and there is no need to cover all possible options, as there may be in a Design-Bid-Build contract. However, this may not be the case for drawings. In a Design-Bid-Build, the Contractor expects that a certain amount of engineering is required before the drawings can be sent to third-party suppliers, such as rebar suppliers or HVAC subcontractors. However, in a Design-Build contract, the Design-Builder often requires that its Designer produce this extra detail, so the drawings can be sent directly to the trade subcontractor or fabricator. Therefore, the level of drawing detail should be clearly defined and agreed upon at an early stage.

Mandatory Requirements

To encourage maximum innovation, Owner mandatory requirements should be kept to an absolute minimum. Where possible, performance rather than prescriptive specifications should be used. One difficult area is the level of quality required. Unless this is clearly specified, the Design-Builder is likely to provide the lowest level that meets the design criteria. Mandatory quality requirements may include cover to reinforcement, minimum concrete strengths, durability requirements, standard of finishes, clearances to vehicle profiles, maintenance requirements, and design life. These types of requirements will encourage innovation and allow the Designer maximum flexibility. Mandatory requirements, such as wall or floor thickness, tunnel diameter, or even the number of tunnel bores, will stifle innovation and will not provide the Owner with the maximum cost advantages that may be obtained by using a Design-Build form of procurement.

Confirmation of Design Intent

During the construction process, the Owner will need representation on-site to ensure that the mandatory requirements are met and to confirm that agreed-upon quality control procedures are being implemented. However, the Designer should also be represented on-site to confirm that the design is being constructed as intended. The representative should be involved in the process of preparing method statements and work procedures and should be allowed the opportunity, for example, to carry out inspections at key construction points prior to a concrete pour. In the contract documents, the Owner should require the Designer to certify that the construction complies with the design drawings and specifications. If certification of this type is linked to final payments, it will ensure that the Design-Builder provides full access and cooperation to its Designer throughout the construction process.

Communication with Owner's Engineering Consultant

Close, effective communication between the Design-Builder's Designer and the Owner's Engineering Consultant is essential if the schedule is to be maintained. Regular meetings must be conducted, and the Owner's Engineering Consultant should be kept fully informed during the design process. It is essential that the Owner recognizes that by using the Design-Build procurement process, the Owner loses control of design details. This loss is often hard for both the Owner and the Owner's Engineering Consultant to accept. However, if this control is not relinquished by either the specification of unnecessary mandatory requirements or by issuing change orders, then design innovation and cost and schedule gains can be quickly lost.

Quality Control

The Design-Build procurement method is more suited than Design-Bid-Build to the Design-Builder carrying out full quality control and the Owner undertaking a quality assurance or audit role. This requires a change in the traditional contracting methods and, when most effective, relies on self-inspection by the construction superintendents. This change of role is often difficult to assimilate as production pressures take precedence. It may be better to set up an independent quality control section, as was done on the England-to-France Channel Tunnel. The Designer's site representative should be closely involved in the preparation of the quality control plans and should be allowed to insert *hold* points for inspection at certain key points. This is especially important if self-inspection is used, and hold points provide for an independent check without being reliant on the Owner's quality assurance audits.

Claims

As with all contracts, a well-prepared set of contract documents will minimize the potential for claims against the Owner by the Design-Builder. However, in Design-Build, not only do the contract documents require interpretation but also the various design codes. For good reasons, design codes allow engineers significant flexibility and recognize that the Engineer must consider the importance of the structure, the result of failure, and other similar issues. There is, therefore, considerable risk that the Design-Builder's Designer, in trying to provide an economic design, will interpret codes differently than the Owner's Engineering Consultant. In the event that the Owner directs a particular interpretation that results in additional cost, the Designer will need to document and provide code interpretation to support the Design-Builder's claim. In addition, the Designer's site representative will need to be cognizant of the assumptions made during the pre-bid design phase so that differing or changed conditions can be identified.

QUALITY OF PERMANENT AND COMPLETED WORK

As in any contractual situation, the Design-Builder is only obligated to provide the level of quality required by the specifications, and, therefore, the importance of specifying the required level of quality cannot be overemphasized. This is often difficult without stifling innovation, but it is the key to a successful Design-Build contract. For example, a Design-Build specification for a car could specify an engine, four wheels, a steering wheel, four doors, four seats, and air conditioning.

Although the Owner expects a Cadillac, the Design-Builder provides a Jeep that meets the specified requirements. As can be seen, the Owner must specify size of engine or the required performance, the required leg room and usable luggage space, the thickness of the body steel, and so forth if the Owner is not to be disappointed with the end result. Detailed specification of requirements is essential if the quality of the finished Design-Build project is comparable with the quality of a fully specified Design-Bid-Build project.

Once the specifications are prepared, it is important that the Owner refrains from changes or additions, because this can lead to significant cost growth. Also, the cost implication of such changes may be more difficult to ascertain, because there may not be a clearly defined baseline. It is also important that the Owner fully considers key items such as durability and maintenance and operational requirements within these specification requirements. If this is not done, the Owner may be faced with a completed project that meets the mandatory requirements but is not durable or maintainable, or that causes operational problems.

Nonconforming Work

Nonconforming work will no doubt be discovered at some time by either the quality control checks or, more significantly, by the quality assurance activities of the Owner. In the first instance, it is important that the Design-Builder resists the production and cost pressures to proceed with the activity. The Design-Builder's Designer should be informed of the testing that indicated the lack of conformance. It is often possible for the Designer to show that either the test requirements were unnecessarily stringent or that an easy fix can be engineered. Additional testing to satisfy other design criteria could be required to prove the adequacy of the member. It must also, however, be recognized that, as Engineers are asked to produce leaner and meaner designs, these options may not be available and the only recourse will be to remove or rework the item that failed to pass the set criteria. The aim of the Design-Builder must always be to get it right the first time and, if it is wrong, put it right as quickly as possible.

If the nonconformance is uncovered by the quality assurance activities, such as an audit, this will create doubt not only about the item found but perhaps the whole of the Design-Builder's quality control program. If procedures are not followed, it may be impossible to determine whether it was an isolated case or endemic in the Design-Builder's organization. If the latter is suspected, it will bring into question the quality of every work item completed to that point. This must be avoided at all costs because resolving such uncertainty may significantly delay the project and affect the Design-Builder's profitability and credibility.

CONSTRUCTION MEANS AND METHODS, AND SAFETY

Means and Methods

Means and methods in either Design-Bid-Build or Design-Build are considered to be the Contractor's responsibility. They include the following:

- Design and use of temporary works;
- Type of equipment used for the project;
- Type of tunneling method or TBM used;
- Shift working arrangements;
- Logistics within the subsurface works;

- Need for temporary access ramps or shafts;
- Order of carrying out the works;
- Size of the labor force and supervision;
- Control of water and ground movement before installation of the permanent works;
- Use of temporary lighting, power, water, drainage, ventilation, and so forth.

Means and methods are the Contractor's responsibility because it can exercise direct control and use its own experience for the benefit of the project regarding the issues that come under that heading. Will the approach to means and methods for Design-Build be different from the approach for Design-Bid-Build? In general, it will be similar because means and methods must always be the Contractor's responsibility. However, three additional comments must be made:

1. If an Owner and its Engineering Consultant wish to prescribe or limit the Design-Builder's discretion for what they consider to be sound reasons—such as a firm view on acceptable risk, a mandated time scale, physical restraints, or bad experience with some methods—then they should do so. This, includes, for example, the generic form of tunneling machine or the tunneling technique. Examples of this in recent years are the St. Clair River Tunnel at the U.S.–Canada border; the Toronto Metro (Toronto Transit Commission) in Toronto, Ontario; Sound Transit in Seattle, Washington; the Rio Subterraneo project in Buenos Aires, Argentina; and the Thames Water Ring Main Project in London, England. On all these projects, the Owner and its Engineering Consultant mandated the use of earth pressure balance machines. Although all these projects were Design-Bid-Build, the same principle would apply to Design-Build projects in similar circumstances.

2. Even if an Owner and its Engineering Consultant wish to dictate a particular type of equipment, they should resist the temptation to include an entirely prescriptive specification. The specification should be performance-related with, perhaps, some basic minimum requirements included. This will allow the Contractor or Design-Builder to apply its own experience without being limited by the specifications, but at the same time will allow the Owner to include what it believes to be an absolutely basic minimum. This subject has caused much debate and will continue to do so.

3. In Design-Build, the Design-Builder has the greater ability than in Design-Bid-Build to tailor the design to suit particular equipment that it owns or methods that it is experienced with to provide a financial advantage.

It should also be noted that on a number of recent projects, the Owner, aided by its Engineering Consultant, has made a direct purchase of the tunneling equipment. In effect, this was done to improve the schedule or to avoid the Owner purchasing the same equipment many times over on projects composed of many repetitive contracts.

Design-Build is a more complex delivery method than Design-Bid-Build, which entails clearer interfaces. Owners use Design-Build in order to obtain a number of perceived advantages, relating mainly to schedule, cost, and risk reduction, plus quality and safety improvements. As it relates to means and methods, the contract must be written so the following five basic aims are achieved.

1. Schedule. As long as the contract negotiations and the bid period have not been allowed to drag on, the construction start date on a Design-Build project should be significantly earlier than on a Design-Bid-Build project. In order to capitalize on this early start, the project must be fast-tracked. The Channel Tunnel is a good example of this fast-tracking, where schedule-critical items such as bored tunneling started before the design or even the location of major elements such as the undersea crossovers were determined. The provisions within the contract framework and the basic specifications and design parameters must encourage this fast-track process. Some of the matters to be addressed are as follows:

- **Design approvals**. The design review and approval procedure has to be geared to cause minimum delay to construction. It is here that Owners and their Engineering Consultants must avoid any temptation to redesign the Design-Builder's designs. On the Copenhagen Metro project in Denmark, arguments between the Owner and the Design-Builder have been based on claimed design approval delay.

- **Authority approvals**. Local authorities are not usually a party to the contract, but they do have strong powers of approval. It is clearly important that the Owner should obtain most of these approvals in advance. Although the Owner is not able to gain advance approval for some of the Design-Builder's means and methods, such as temporary access shafts, ground treatment from the surface, and other temporary works, it can anticipate some of these needs by having early constructability analyses and early discussions with pre-qualified Contractors. In addition, an Owner can aid the approval process by good communications with the local authorities and other third parties throughout the life of the project.

- **Permits.** Permits and rights-of-way (ROWs) for temporary works, diversions, and so forth can cause delays to contracts if not properly handled. Ideally, the Owner will obtain the majority of permits and ROWs before letting the Design-Build contract. However, some of the permits and ROWs will depend on the Design-Builder's means and methods and can, therefore, only be obtained when the details are known. If not properly handled, this can cause major disruption to the approval procedures and delay the project.

2. Cost Savings. Potential cost advantages with the use of Design-Build can be obtained from three sources:

1. Time savings due to fast-tracking of the project.

2. The ability of the Design-Builder to carry out the economic design to suit its own well-tried means and methods. The Design-Builder can, in effect, put together all the advantages of economic design, constructability analysis, and value engineering in order to arrive at the most economic solution.

3. Early use of the facility, allowing early return on capital invested.

3. Risk Reduction. Regardless of whether a project is Design-Bid-Build or Design-Build, the risk analysis process should be initiated as early as possible. On a Design-Build project, risk evaluation is a live process throughout the life of the project. Construction risk is overwhelmingly influenced by means and methods, which reflect the design, so a procedure is needed to ensure that all the contracted parties play their part in the risk analysis process. Of course, there is some danger that this can be the subject of contractual contention, and, in fact, this happened on the Copenhagen Metro. In some instances, agreement can be made in advance that joint participation in risk analysis can be kept out of the contractual arena.

4. Quality. An approach similar to Design-Bid-Build can be taken with respect to quality. By starting the evolution of quality plans at the design stage and marrying these with the Design-Builder's means and methods, the relationship between design and construction quality can be closer at the earliest possible stage of design. The view that with a Design-Build contract the

Design-Builder can do all its own inspection and quality checking without any oversight from the Owner has several dangers:

- With no independent inspection on behalf of the Owner, when it comes to contractual disputes there are only one-sided records;
- It is difficult for the Design-Builder's inspection team to exercise real control over the work; and
- With subsurface construction, mistakes are often difficult to correct.

It is absolutely vital that the quality procedures reflect the type of contract and allow both the Owner and Design-Builder to achieve their quality objectives.

5. Safety. Like quality, the approach to safety should be similar to that on Design-Bid-Build. Because the designs fit with the Design-Builder's means and methods from a safety standpoint, there is a real opportunity to take a unified approach to design and construction safety. This will involve the design team being in constant liaison with the construction team and carrying out analysis for continuing risk, constructability, and safety.

To summarize, in terms of means and methods, the two delivery methods, Design-Bid-Build and Design-Build, are very similar. Design-Build provides a number of opportunities

- To create a closer relationship between design and means and methods; and
- To address the fundamentals of schedule, cost, quality, safety, and environment in a more coordinated manner throughout the life of the project.

In order to take advantage of these opportunities, the Design-Builder, aided by the Owner, must facilitate the fast-tracking of the project by establishing efficient procedures for design review and other third-party approvals and permits. The Owner needs to give the Design-Builder the freedom to adapt designs and means and methods to suit its own expertise, providing that the Design-Builder can deliver a high quality and safe project on time and within budget without unnecessary impact on the environment or the local community.

THIRD-PARTY AND OTHER PUBLIC CONSIDERATIONS

Subsurface projects, especially in urban and populated areas, require significant efforts for mitigating construction impacts. Although the public wants as little disturbance as possible, construction with no impact is not possible. However, prior to construction, a project should be "sold" to the public and concerned stakeholders by including commitments to mitigate negative construction impacts, which can be documented by the environmental impact statement, by memoranda of understanding with abutters, and through other agreements.

Impacts during subsurface construction can include

- Damage to existing buildings, utilities, and facilities along the right-of-way due to lateral movement of excavation support walls and consolidation from construction dewatering;
- Damage to existing facilities from construction vibration;
- Construction noise;
- Dust and impacts to air quality; and
- Impacts to surface vehicle and pedestrian traffic due to construction staging.

These impacts are directly related to means and methods. For example, when constructing a slurry wall, the choice of a clamshell bucket with chisel for hard material will result in more vibration than use of a hydro mill–type of machine. A second point to note is that construction methods that influence impacts have been increasingly defined and specified in construction contracts. Environmental commitments to avoid negative impacts have led to specified rules requiring that standards be met during construction. The analysis and specification happens before any shovel hits the ground.

This has led to a dilemma in Design-Bid-Build contracts for subsurface construction. Because the design documents must, to an extent, specify and control activities considered more in the realm of the Contractor's means and methods, the requirements lead to a further blurring of the line between design and construction.

In Design-Build, the Designer and Contractor are on the same team. The activities of analysis, design, and construction are not as contractually separate as they are on a Design-Bid-Build contract. However, this leads to new dilemmas for mitigation of construction impact. Regardless of the contract delivery system, the Owner is still responsible for mitigating negative construction impacts. Requirements for mitigation need to be defined long before construction begins, working with the same cast of characters of responsible government agencies, authorities, and local stakeholders. But prior to bringing the Design-Builder on board, there is no way to test how the detailed design will lead to construction impacts, because there is no detailed design.

The solution is to develop a fairly detailed performance specification that defines what the final design must show and what standards the construction must satisfy and be measured against. Although a performance specification defines impacts such as excavation-induced movement, noise, dust, and others, enough background engineering work must be done to verify that it is possible to achieve the performance. A Design-Build specification may state that no additional noise shall be allowed during construction. This looks good on paper, and reviewing agencies and the local community may be cheered by these requirements, but such a specification would be impossible to satisfy and could not be supported by analysis.

CONSTRUCTION STAGING ISSUES, ADJACENT CONTRACTS

Many subsurface projects are subdivided into separate construction contracts:

- The mainline excavation/structural work may be placed in one contract, the finishes work in another, and the systems/mechanical/electrical work in a third contract. Cut-and-cover tunnels on the Central Artery/Tunnel Project in Boston, Massachusetts, were let using this approach.

- Construction contracts may be let based on schedule requirements and needs. For example, on the Shot Tower Station project, part of the northeast extension of the Baltimore Metro (in Maryland), the slurry walls, excavation, and base slab were part of the first contract. The second contract included roof, finishes, and other elements. On the NERL light rail subway extension in Newark, New Jersey, the first contract included the cut-and-cover tunnel section at Mulberry Street, but not the transition "boat" section.

- Contracts may be let based on available funding. Also on the NERL project, subsequent sections of the tunnel connection to the Newark airport were planned to be awarded at later dates.

- Often subsurface construction contracts are so large that they are subdivided into separate pieces to help make the construction more manageable and to encourage more bidders.

In addition, tunnel projects may require coordination with other external projects. Tunneling is often done in urban areas and may need to work in and around building projects, utility work, and transportation improvement projects. For example, portions of the Central Artery/Tunnel Project in Boston had to be coordinated with the Amtrak rail electrification, Lincoln Street (a deep excavation/multi-story skyscraper), and the MBTA North Station subway and development. Each external project imposed unique scheduling and space constraints.

Overall, subsurface projects require significant coordination of construction contracts, both internal and external. In the Design-Bid-Build approach, coordination is handled as follows:

- Specifications can include *access restraints*, periods during which the Contractor is prohibited from working on a particular area or structure. For example, if an adjacent contract is to provide access to a work zone, but that access is not scheduled to be available until four months after notice to proceed on the subject contract, the specifications can include an access restraint to that work zone.

- The design drawings can include specific notes and description of structures and facilities provided by others, along with the responsibilities of the subject contract. For example, if an earlier contract provides excavation support walls, the subject contract will need to include baseline documents about the walls and their location, their capacity, how they are to be incorporated in the current construction, and so forth.

For Design-Build contracts, the requirements of contract coordination pose a dilemma. In Design-Bid-Build contracts, the requirements can be better defined prior to letting the contract, because the design is better developed. For Design-Build contracts, it is still necessary to define what is expected of the Design-Builder for contract coordination, but this definition must be based on a conceptual or pre-preliminary level of design. This can be achieved as follows:

- Define coordination requirements and responsibilities, and require the Design-Builder to be responsible for implementing them. This solves the technical problem of less design definition but effectively adds risk to the Design-Builder. The result can be higher bid prices to cover the risk.

- Analyze or design certain elements to a higher level of completion to reduce risk and thus improve bid prices. For example, in subsurface projects, a high level of geotechnical investigation and analysis can be performed prior to the Design-Builder coming on board. This early information helps to reduce risk caused by subsurface obstructions, contamination, and other problems unique to tunneling.

- Develop the contract such that the Owner takes responsibility for coordination of certain elements. In many cases, the Owner may be in a position to assume this responsibility at a lower risk (and thus cost) than the Design-Builder. For example, some utility relocation work on the University Light Rail Transit Project in Salt Lake City, Utah, was separated from the main Design-Build contract, and the utility companies were paid for required work (once this was established during the Design-Build process) by lump sum. Because the risk to the Design-Builder was effectively reduced, the Design-Builder no longer had to deal with this variable, and the Owner was in a position to deliver needed coordination.

MANAGING PUBLIC RELATIONS

Managing public relations in a construction project largely involves communicating to the public the expected adverse impacts of construction and mitigating them. Adverse impacts can include

- Noise,
- Dust,
- Vibration,
- Reduced air quality,
- Disposal of hazardous materials, and
- Traffic staging and possible disruption.

A subsurface construction project must deal with these issues and some special concerns peculiar to tunneling and excavation:

- Excavation can lead to ground subsidence. In a cut-and-cover tunnel, support of excavation walls tends to deflect inward. In a bored tunnel, the ground can subside above and around the cut, particularly as the face progresses. The ground movement can impact adjacent buildings, utilities, and other structures. These impacts can be of particular concern for historic older buildings.

- Tunneling and subsurface construction may require temporary dewatering and pressure relief. The goal is to dewater only in the cut or at the construction face, but limiting impacts to the surrounding substrata may be difficult. Dewatering in the adjacent areas can result in consolidation, again a problem for existing structures. Also, it is important that the existing water table be maintained beyond the cut to avoid exposing structures founded on timber piles. This applied to the Copenhagen Metro project and will apply to the Citytunneln project in Malmö, Sweden.

Dealing with the public in managing and mitigating construction impacts requires a combination of good planning, specification, control, and communication. Considering the requirements of Design-Build contracts, there are advantages and disadvantages to how a Design-Build project deals with public relations in comparison to a Design-Bid-Build contract. In the more traditional Design-Bid-Build contract structure, there is a more defined, separate design phase. During this phase, specific construction sequences and work zones are at least schematically defined; a performance specification is developed for the Contractor's means and methods to address all adverse impacts; and, ideally, the public is invited to participate in the definition and specification. For example, noise specification limits are defined (what is a decibel, how it impacts residents, how it drops with distance, etc.). The public participates in the discussion of the tradeoff for setting reasonable limits that allow construction while fairly mitigating negative impacts.

A disadvantage of the Design-Bid-Build approach for this issue is that all definition, discussion, and specification of construction methods occur during the design phase based on assumptions and predictions of the Constructor's means and methods. When construction actually begins, the reasonableness of the specifications is tested against other requirements such as constructability, construction schedule, and the Contractor's means and methods and proposed redesign. The issue, in a nutshell, is that the only way to have absolutely no construction impacts at all is to build nothing.

In a Design-Build contract, the design is performed with more input from those responsible for means and methods, so it is possible to better model and evaluate adverse construction impacts. The disadvantage is that the design documents, from which the evaluations are made, may not be as complete and easily evaluated as in Design-Bid-Build. For example, design of one section of the project may be complete, and construction may begin prior to design of another section. Both sections may be important for evaluation of a particular construction impact such as traffic staging, but it may not be possible to do the same type of evaluation, because analytical information is not complete. However, the public still must be informed and still wants to be assured that impacts are minimized.

This communication can be achieved by tight performance specifications and in some cases development of a greater level of advance design detail than would be normally done in a Design-Build contract. Considering the traffic-staging example, if the critical path requires one section of the tunnel to be designed and built immediately but is not required on an adjacent section, what if a critical surface traffic route passes through both sections? To demonstrate to the public that traffic needs will be met during construction, it may be necessary to advance both tunnel sections to a level of design complete enough to establish how the traffic route will be maintained.

To communicate with the public, Design-Build teams should use methods similar to those used on Design-Bid-Build contracts. These include

- Public meetings;
- Interaction with agencies and affected organizations; and
- Use of new communication tools, such as a project web site.

Because the design schedule tends to be more compressed than in Design-Bid-Build, to meet the goals of informing the public and getting buy-in for construction plans, the frequency and intensity of communication needs to be greater.

COMMUNICATION AND PARTNERING

Clear and effective communications are particularly important in Design-Build because the Owner and the Owner's Consultants are not directly involved in the management of the work, and any miscommunication may not be noted and corrected until considerable time and expense has been incurred. While informal communication on all levels is to be encouraged, formal communications should be channeled through the Owner's Construction Manager, so they are properly coordinated and documented. It may be advantageous to draw up a communication flow diagram, if only as a means of demonstrating to individual staff or departments within the Owner's organization, who wish to have their own direct contact regarding certain elements of the work, how this will quickly lead to unmanageable complexity.

Clearly, communications between all parties is essential in a Design-Build project. The most essential rule, though, is that the Owner's Engineering Consultant must not be tempted to redesign the Design-Builder's designs, thereby causing unnecessary delay to the project.

The expectation in a Design-Build project is that, ideally, processes will be streamlined. The adversarial relationships that can develop in Design-Bid-Build are not present because there are fewer separate organizations. The flip side is that communication is essential among all parties and stakeholders. Having a Design-Build team as one entity requires that all components of this organization be in contact and clued in to the process. The challenges can be magnified for a subsurface Design-Build project, because subsurface work may include unforeseen conditions, obstructions, and construction challenges not present in aboveground construction.

Team Members

Team members in the Design-Build process include the Design-Builder and its Designer; the Owner and its Owner's Engineering Consultant; and outside stakeholders, such as utility companies and abutters. Good communication with outside stakeholders is particularly important for Design-Build delivery. In a Design-Bid-Build project, it is typically the Engineer's responsibility to communicate and coordinate with stakeholders. For example, the impacts on utilities are coordinated and specified as part of the design. However, in Design-Build, the design process is streamlined, with construction often beginning before design is complete. Frequent communication is needed with outside stakeholders so their concerns are addressed during the rapid iteration. In some Design-Build projects, representatives of stakeholders are actually incorporated into the team and empowered to address the stakeholders' interests and make decisions during design and construction to ensure their interests are met.

Partnering

In recent years a formal partnering process has developed as a way of dealing with adversarial relationships. On a traditional Design-Bid-Build project, requirements for formal partnering sessions are specified in the bid documents. Once the Contractor is on board, partnering meetings are held at different times during the project (startup, 50% completion, etc.). The meetings may last from a day or two to a week and are often held off-site. At these meetings, representatives of the Contractor, Owner, Engineer, and other parties participate in team-building exercises and develop conflict-resolution procedures. The project participants agree to the procedures via sign-off on formal documents. The overall intent is that, by agreeing to objective conflict resolution procedures and through the benefit of face-to-face procedures, conflicts can be more easily managed and vilification of one side by another can be avoided.

The idea of partnering still has application in Design-Build projects. Conflict in design and construction is inevitable, so some form of conflict resolution is required. One difference in Design-Build is that the tension between design and construction is not expressed by different organizations but is part of one organization responsible for the final design and construction. The principles of partnering to address this tension may not be as formally applied as with Design-Bid-Build. However, it is still essential that Designers and Contractors in a Design-Build project see themselves as partners and not adversaries. Whether this is done by formal off-site partnering meetings and procedures or by less formal internal methods, it is still necessary if the project is to succeed.

The Owner's organization chart from the University Light Rail Transit Project in Salt Lake City exemplifies the degree of coordination between design and construction expected in a Design-Build project. It refers to a design compliance manager, a separate branch of the organization responsible to enforce design quality issues. Some positions are labeled as design auditors, who were responsible for conducting frequent quality design audits.

Because construction projects have problems and conflicts, Design-Build is not a panacea. Alternative dispute resolution such as dispute review boards should be specified. Instead of a separate Designer and Contractor in the traditional Design-Bid-Build, the board deals with a Designer-Contractor, but the general theme and approach are similar. The dispute review board is established as a formal agreement between the Owner and the Design-Builder. Representatives are selected, and procedures are followed through a specified, contractual procedure.

Meetings

As in Design-Bid-Build, the business of Design-Build project communication is largely handled by project meetings. The frequency and subject of meetings may be impacted by the compressed design schedule. If construction begins before design is complete, required adjustments to design are discussed in meetings. Some Design-Build projects such as the Utah light rail transit extension scheduled regular "change" meetings to address these topics. In addition to Design-Build members, other stakeholders were invited to participate at these regularly scheduled meetings. The meetings provided a forum for continuous dialogue, communication, and resolution of issues as they came up.

In Design-Bid-Build, quality is often thought to be the priority of the Owner, while cost and schedule are higher priorities of the Contractor. Ideally, the interface between the Owner and its consultants and Contractor leads to the processing and satisfactory resolution of quality, cost, and schedule. In Design-Build, because one organization is directly responsible for quality, cost, and schedule, some Owners have been concerned that this arrangement ends up placing higher priorities on cost and schedule, with quality getting the short end of the stick. Design-Build projects have addressed this concern, in part, by programming discussion of and responsibilities for quality into the communication process. Therefore, regular "quality" meetings, with actions, responsibilities, and documentation, are built into the process. If the participants are empowered to react to issues and deal with them, these meetings are effective. Because quality cannot be discussed in a cost–schedule vacuum, the inevitable tension between quality, cost, and schedule must be addressed.

For subsurface Design-Build, safety is a special concern. Subsurface projects with hazards that include confined spaces, excavation support, hazardous materials, and blasting have additional safety challenges. As part of project communication, Design-Build projects schedule regular safety meetings and safety walks. Because of the rapid design and construction iteration, for the communication to be effective, all participants must be connected to the developing information.

The Design-Build approach should conceivably lead to less paperwork and more streamlined documentation, because there are fewer formal entities with which to correspond. However, this is not necessarily the case. Owners not familiar with Design-Build arrangements can develop procedures that, in terms of documentation, are really geared toward Design-Bid-Build. The result can be more paperwork, not less, with some duplication of functions within the organization. Also, the speed of design iteration requires careful documentation by the Design-Builder. The frequent meetings require frequent and thorough minutes, with assigned responsibilities and follow-up. In some ways, the demands for documentation in Design-Build are greater.

Ultimately, communication comes down to understanding issues and dealing with each other's problems. This is not unique to Design-Build, Design-Bid-Build, or any other form of project delivery, for that matter. The special challenges in subsurface construction—obstructions, "mystery" buried utilities, and many others—can only be dealt with if participants know about the issues and are empowered to deal with them. Because Design-Build can compress the design schedule and shorten reaction time, an effective project communication scheme needs to be able to react to the challenges.

REVIEWS OF SHOP DRAWINGS, OTHER SUBMITTALS, AND DESIGN-BUILDER'S DETAILED DESIGNS

Review of the Design-Builder's design, drawings, and related documents is critical to ensuring that the resulting facility meets the Owner's requirements. The reference design provided by the Owner is a preliminary one that defines the requirements for the facility in terms of space and functional requirements, and possibly appearance, and third-party requirements or agreements. In some cases, the reference design defines certain drawings and specifications, or parts thereof, as mandatory, though this can cause some confusion for the Design-Builder determining which parts of such drawings are mandatory and which are merely illustrative and included on the drawing as context for the mandatory features. Inevitably, there is some degree of interpretation of these requirements that needs to be verified by review of the Design-Builder's detailed designs. The Owner's review must resist the temptation to change or impose unnecessary additional requirements during the review process; such changes would likely delay and increase the cost of the project.

Design-Builder's Drawings and Specifications

The Design-Builder is required to produce the detailed design and submit it for the Owner's review in the form of drawings and technical specifications. The detailed design drawings and specifications should be fully detailed to similar standards as would apply to bid drawings for a conventional Design-Bid-Build contract and should be accompanied by the pertinent calculations.

The necessity to submit working or shop drawings for review depends on the level of quality control being exercised by the Owner. If quality control is primarily the Design-Builder's responsibility, the Owner's Construction Management Consultant can perform quality control oversight and spot checks using technically qualified staff who have the ability to determine whether the work conforms to the intent of the detailed design drawings. However, depending on the level of detail, shop drawings may still be needed as as-builts.

Early Drawing List

In order to ensure effective management of the project and to allow the Owner's Consultants to plan their own work, the Design-Builder should be required to provide within 30 days of notice to proceed a schedule of design submittals, giving the list of drawings and specification sections that will be included in each. The schedule, which will be reviewed by the Owner's Consultants to ensure that it is complete and that it supports the construction schedule, should be incorporated in the Design-Builder's overall schedule.

Early Submittal List

A schedule of other submittals should also be provided within 30 days of notice to proceed. The number and extent of submittals that the Owner's Consultants review will depend on the allocation of responsibility for quality control and on the Owner's need to have full documentation of the work. For clarity of all parties, the contract should include lists of submittals required in the form of contract data requirements lists (CDRLs) for each section of the technical specifications. For the internal use of the Owner's Consultants, the master submittals list based on the CDRLs should indicate primary and secondary responsibilities for submittal review.

Agree Procedure

Procedures for transmitting, receiving, reviewing, tracking, and closing submittals should be established before the start of construction. There is no significant difference in the procedures to be adopted for Design-Build contracts, but the clarity and implementation of the procedures may be more critical because delay and misunderstandings, particularly concerning the Design-Builder's submittals of final designs, may have far-reaching consequences. The procedures concerning submittals should be equitable and not impose an unreasonably tight time scale on the Contractor with regard to resubmittals, and not give the Owner and its consultants an unreasonably long time in which to respond.

Nontechnical Submittals

A number of nontechnical submittals are also required. For example, all parties involved share responsibility for construction safety, and the Owner has a responsibility to check that the Design-Builder has the legally and contractually required safety provisions in place. Typically, there are also requirements for the Design-Builder's management personnel and for certain key operatives to be suitably qualified and experienced. There may also be requirements, depending on the jurisdiction or funding source, to employ a diverse workforce, certain disadvantaged businesses, or to pay minimum wages. Other nontechnical issues include bonding, insurance, and lien releases. Some of these requirements are covered in the pre-qualification process, and the Owner should verify that the Design-Builder's pre-qualification statements are being fulfilled.

Design Review

The contract should make clear what level of design submittal is required and to what level of review it will be subjected. Normally, the Owner will require design calculations at the level of detail necessary to support any future changes to the structural elements or to investigate any problems that may arise. However, it is not essential to submit the design calculations in full but only to summarize the design in the form of a report in a standardized format. The Design-Builder's final design should, of course, be prepared by suitably qualified professional engineers, subjected to internal checks and reviews, and sealed by a professional engineer before being submitted for review by the Owner's Engineering Consultants. As such, the Owner's Engineering Consultants do not need to check the design in detail for technical accuracy but should focus on the contract's conformance to the reference design, the design criteria, and the design intent and coordination with work performed under other contracts.

Shop Drawing Review

Even though the Contractor is preparing final design drawings, these will be to the same level of detail as the contract drawings prepared by the Owner's Engineer in a conventional Design-Bid-Build contract. Working or shop drawings will still need to be prepared for the use and guidance of workers in the field. Approval of shop drawings should not relieve the Design-Builder of the obligation to comply with the final design drawings that it has prepared, submitted, and had approved as part of the design review and approval process. However, if changes to the final design are found necessary or desirable during the preparation of working drawings, the Design-Builder is at liberty to revise and resubmit.

Construction Work Plans

Construction work plans should be required for each major element or phase of construction, the work sequence should be set out, and the construction methods and equipment should be identified. Each construction work plan should include a job hazard analysis, and a critical part of the plan should be the safety and quality control procedures specific to the work covered, so they supplement the Design-Builder's overall safety and quality procedures. The Owner's Construction Management Consultant should review the construction work plan, with input from the Owner's staff or consultants for safety and quality.

Materials Review

Formal submittal of material samples should only apply to cases where appearance is critical and cannot be adequately specified. It should be noted, however, that if the Owner wishes to select or adjust the appearance of materials, this needs to be made clear in the contract, with times for review and resubmittal specified, so the Design-Builder can make appropriate allowance in the schedule and avoid claims for delay or additional direct costs.

Document Control

Document control for submittals is a subset of the overall configuration management and document control procedures. It is important to track submittals to ensure that they are received, reviewed, if necessary resubmitted, and approved so as not to delay construction. Submittals must be filed and made available to field staff to facilitate oversight of the work, and must be safely kept as part of the overall contract records.

Quality Control

Quality control of submittal preparation should be the Design-Builder's responsibility and should fall within its overall quality plan. Some elements of quality control that pertain to submittals may be specified in the contract, particularly those concerning preparation of the Design-Builder's final design.

Quality Assurance

The results of the Design-Builder's own quality assurance audits should be submitted to the Owner, and the Owner or the Owner's Consultants should also conduct independent audits to verify compliance.

RESPONSIBILITIES OF OWNER'S ENGINEERING CONSULTANTS

Although quality assurance methods, involving review of and audit of compliance with procedures, are a critical component of ensuring quality work, it is also important that qualified and experienced personnel directly observe the work on the Owner's behalf to ensure that the paperwork lines up with physical reality. The Owner's Engineering Consultants, particularly the Construction Management Consultant, bear the main responsibility for this, though on a Design-Build contract the Owner's Engineering Consultant should also be on-site to verify that the original reference design intent is being followed as it is interpreted and implemented in the Design-Builder's final design.

Verify Design Assumptions

With any subsurface construction, a range of design parameters is developed from the site investigation, and either the works are designed for the worst credible conditions, or an observational approach, such as the New Austrian Tunneling Method, is adopted. In either case, it is important to verify during construction that the design assumptions are valid and are being met. Most tunneling contracts include monitoring to verify performance in accordance with design assumptions, and, in the case of an observational method, detailed procedures are laid down that give predicted ranges of monitoring results and the steps to be taken if unpredicted events occur. For a Design-Build contract, the Design-Builder is responsible for the final design and should also be responsible for verification of design assumptions as construction proceeds. However, as demonstrated by some widely publicized failures, the consequences of failing to properly verify design assumptions and performance are such that the Owner's Engineering Consultant should closely oversee the monitoring and interpretation of the results. This requires staff experienced in the type of work being performed and is additional to the quality assurance function.

Ensure Design Intent Is Being Met

Another important role of the Owner's Engineering Consultant is to ensure that the design intent is being met. The Owner's reference design should include all the design requirements (in terms of function, appearance, compatibility, third-party agreements, etc.) that are important to the Owner but allow the Design-Builder the maximum freedom within these overriding constraints. However, the Design-Builder will inevitably make choices that were not anticipated in the reference design and that do not meet the design intent. It is important to recognize such issues at the earliest possible moment, so that the design requirements can be clarified and, if necessary, a change issued.

Act as Independent Source of Information

Although Design-Build gives the opportunity to adopt measures to reduce claims, disagreements will inevitably arise over interpretation of the contract or over differing site conditions. Even in the majority of cases where the Owner agrees that a change is justified, there may be disagreements on the time and cost impacts of the change. In such cases, the Owner must have a source of factual information that is independent of the Design-Builder in order to evaluate such issues. Even if the Design-Builder has the main responsibility for quality control, the Owner's Construction Management Consultant should keep daily records of activities on-site, including workers, equipment, conditions, and the project progress.

Monitor Design-Builder Performance

In addition to being equipped to deal with Design-Builder claims, monitoring the Design-Builder's performance and work progress has a number of other purposes. Assessing progress payments to the Design-Builder depends on knowledge of the work performed. A proactive approach to construction delays or differing site conditions requires knowledge of the issues, such as determining whether a delay occurred because of a learning curve or whether an ongoing problem that will lead to a seemingly minor delay becomes a major issue. It is in the Owner's interests to have the Owner's Engineering Consultants directly monitor the Design-Builder's performance to independently verify the accuracy and completeness of the Design-Builder's reporting.

Do Not Abdicate Responsibility

One of the perceived advantages of Design-Build is that it places the responsibility for both the final design and the construction with the Design-Builder, giving it the control and responsibility for more of the construction risk than with Design-Bid-Build and, in particular, relieves the Owner of the responsibility for coordination between final design and construction. However, the Owner should not abdicate responsibility for protecting its own interests, either directly or through its Engineering and Construction Management Consultants. It is important to remember that the Design-Builder's priorities may be different from the Owner's, and the Design-Builder may be willing to make decisions or take risks that are not in the Owner's best interests. For example, the financial penalties (liquidated damages) for late completion may be insufficient to motivate the Design-Builder to mobilize additional resources to mitigate delay and at the same time be insufficient to compensate the Owner for the true delay cost if the contract falls on the critical path of the overall project. It is important, therefore, that Design-Build, while transferring some well-defined responsibilities and associated risks to the Design-Builder, is not seen as a hands-off option where the Design-Builder is expected to deliver a completed project without Owner intervention. Overall, the roles and responsibilities of the Owner and the Owner's

Engineering Consultants for a Design-Build contract are not much changed from conventional Design-Bid-Build.

CONCLUSION AND RECOMMENDATIONS

It is clearly important for all the contracting parties in a Design-Build project to recognize the differences in their roles compared to a Design-Bid-Build contract. Owners should not use Design-Build as an excuse to off-load all the risk onto the Design-Builder. Risk allocation on an equitable basis is to both parties' advantage. Communications during construction within a Design-Build contract becomes even more important, and this must be recognized within the framework of the contract. Independent inspection by the Owner's Consultants remains a vital function and should not be replaced by Design-Builder self-inspection. The Owner needs to retain an independent auditing role.

The two main advantages of Design-Build are schedule reduction and the marriage between design, and means and methods. These advantages can easily be destroyed by overzealous Owner's Engineering Consultants, Owners' reluctance to release control, and lack of advance planning on permits. Design-Builders must be prepared to give a degree of independence to their own Engineers, or otherwise the ability to match the design intent can be difficult to achieve.

Design-Build is the current flavor of the month. If the industry wishes it to become a permanent taste, then all parties must use it in an equitable and proactive manner for the benefit of the contracting parties and third parties.

This chapter covered many issues relating to the construction phase that are equally applicable to Design-Bid-Build or to aboveground work. However, there are five main considerations that apply with special force to Design-Build subsurface contracts. These are intended to focus on special aspects of Design-Build construction and help determine whether Design-Build is the correct approach for a particular project and identify factors crucial to its success.

Recommendation 10-1: For a typical Design-Build contract, the usual principle that the Contractor takes responsibility only for those risks over which it has some control still applies. Specifically for subsurface work, "The Owner owns the ground" and differing site conditions clauses apply. The Geotechnical Baseline Report will set out the conditions to be expected and the less likely but still possible conditions that the Contractor may have to deal with and for which it is compensated according to the contract.

Recommendation 10-2: During the construction phase, the risk register should be maintained as an active document, regularly updated to identify new risks and re-assess existing ones, and to actively manage risks through the construction and commissioning phases and successful project completion.

Recommendation 10-3: When disputes occur, effective formal and facilitated partnering can lead the parties to resolve the dispute in a constructive manner and minimize the impact on progress of the work. The Design-Build contract should include an alternative dispute resolution provision for disputes that cannot be resolved between the parties, with the aim of speedy resolution through a fair and transparent process that all parties respect and are likely to accept.

Recommendation 10-4: To avoid lengthy design submittal reviews, rework by the Design-Builder's Designer, and disputes and delay claims, design workshops should be conducted in which the Owner's Consultants explain the development of the reference design and share their knowledge of the ground conditions, while responding to the Designer's proposed approach to the design and design innovations. Initial design workshops should cover the main areas of design, and additional workshops should be instigated by the Owner's Engineering Consultant when review of design submittals shows differences in approach or apparent misunderstandings of requirements.

__Recommendation 10-5:__ The Design-Builder's responsibility for quality control should be accompanied by responsibility for quality control—overseen by appropriate quality assurance audits. For underground construction, it is in the Owner's best interests to have its own forces or consultants perform some level of direct inspection and an independent laboratory perform verification testing. The quality assurance audits should include direct observation of the inspection process and independent review of the data collected.

CHAPTER 11

Insurance Coverage Issues

Cesare J. Mitrani
Executive V.P., Willis Construction Practice, Dallas, Texas

David Grigg
Senior V.P. –National Director, Willis Construction Practice, New York, N.Y.

Mike Anderson
Senior V.P.–National Director, Willis Construction Practice, Radnor, Pa.

INSURANCE, SURETY, AND RISK MANAGEMENT CHALLENGES

As Design-Build entities are discovering, there are formidable challenges to developing a cohesive insurance, surety, and risk management program to address the exposures from Design-Build work. This issue is heightened by the increased use of turnkey project delivery, where the Owner seeks one entity to provide a seamless franchise in order to outsource a host of services. Design-Builders are being called upon to provide project financing, site selection purchase, operation and maintenance, and possibly an equity ownership interest, in addition to architecture, engineering, and construction services.

Liabilities and Damages

In these expanded roles, the Design-Builder may be subject to an array of liability and potential damages. The source of this responsibility can be attributed to contractual liability (e.g., indemnity warranty, liquidated damages, performance guarantees) and various types of tort liability that derive from the broadened scope and role of the Design-Builder in the design and construction process. The theories of tort liability include

- **Negligence**. The Design-Builder fails to carry out its functions in a manner consistent with normal and accepted standards of practice.

- **Strict liability/joint and several liability**. Design-Builders that engage in hazardous activities can be held liable for all injuries that those activities cause, even if the Design-Builder was reasonable in providing the services. Under the theory of joint and several liability, if there is more than one defendant liable for the injuries, the injured party may collect damages from any one, or all, of the defendants.

- **Statutory liability**. Federal and state liability statutes can impose strict liability without regard to degree of fault. This is especially true for Design-Build environmental remediation projects where the parties can be held liable to CERCLA (Comprehensive Environmental Response, Compensation, and Liability Act of 1980, better known as Superfund) and other state environmental statutes.

Operational Exposures

With the Design-Builder providing a broad scope of services, an inherent interrelation of operational exposures are associated with the work, which can include, for example,

- Faulty design;
- Faulty construction workmanship;
- Environmental damage;
- Injury to workers and third parties;
- Damage to property of Owner, Contractor, supplier, and third parties; and
- Performance deficiencies.

Through contractual obligations and tort liability, the Design-Builder or its subcontractors/subconsultants can be held responsible for these injuries and damages.

INSURANCE AS A RISK TRANSFER, RISK FINANCING TOOL

Effective risk management requires an understanding of the available insurance products and the ability of insurance to absorb certain risks that are inherent in a Design-Build project. Thus, the effectiveness of insurance to transfer risks depends on both the Design-Build contract's provisions relative to the procurement of insurance and the operative terms of the specified insurance policies.

One of the significant issues in Design-Build is risk transfer. In an article about risks in subsurface construction projects (Brierley and Cavan 1987), five major risks that needed to be managed in order to have a successful outcome for the project were identified:

1. Third-party impacts
2. Differing site conditions
3. Design defects
4. Contractor incompetence
5. Contract document

In risk management planning, three kinds of situations make the use of risk transfer appropriate:

1. The risk can be too large for the Design-Builder to retain and be able to achieve its objectives.
2. The Design-Builder has a legal obligation to transfer the risk.
3. Risk transfer is the most efficient management device for the exposure, even though loss retention is feasible and risk transfer is not required by statute or contract.

Contracts of insurance are the principal means for financing losses too large to retain safely and too expensive to prevent or avoid. Insurance then becomes a hedge against risk. By paying a predetermined insurance premium, one can transfer risk and thereby avoid the possibility of large catastrophic losses.

INSURANCE COVERAGE APPLICATIONS

The purchase of insurance and surety is a way to transfer the risk associated with these exposures to an insurer or bonding company. Table 11.1 illustrates this point.

TABLE 11.1   Traditional insurance/surety applications

Coverage	Exposure
Commercial general liability	Third-party bodily injury and property damage, faulty construction workmanship
Workers' compensation and employer's liability	Injury to workers
Professional liability	Faulty design or other professional services
Contractor's pollution liability	Third-party environmental bodily injury and property damage and clean-up
Builder's risk	First-party damage to property of Owner, Contractor, or supplier during construction
Efficacy/contingency risks	Liquidated damages, debt obligations
Surety	Performance and payment obligations of Contractors

Design-Build Insurance Challenges

Although a variety of insurance products are available, the coverage afforded by one policy does not necessarily dovetail with another policy to provide a seamless fit. It is important to review the standard policy exclusions and limitations to determine the true extent of coverage. Many standard exclusions are broad in their application, which can substantially restrict the coverage. The key insurance challenge is to properly dovetail the standard policy exclusions to address the interrelated project-specific exposures on a Design-Build project. To illustrate this point, Table 11.2 outlines ways to treat several standard exclusions contained in professional, general, and pollution liability insurance policies.

Professional Liability. The professional liability exposures associated with Design-Build of subsurface projects primarily arise out of potential design errors or omissions by the Engineering Consultants engaged in the design of the project. It is possible, however, that the Design-Builder may also have professional liability exposure associated with its performance as a construction manager, particularly with respect to any design duties delegated to specialty subcontractors.

Consequently, two distinct approaches address the professional liability risks: by means of an Architect/Engineer (A/E)-based or Contractor's professional liability policy wording. It is common on Design-Build projects to have these policies in place on a project-specific basis, with the limits dedicated to the subject project. Project-specific professional liability is a problematic class of business to underwrite, particularly for subsurface projects such as tunnels. Only a limited number of insurers will provide capacity and insurance coverage for subsurface projects, which are subject to high premium and retention levels (typically $1 million to $10 million).

If an A/E wording is utilized, coverage is afforded to the design team for acts, errors, or omissions in the provision of design services. The prime designer and its subconsultants will typically be identified as the named insureds. Traditionally, the Design-Builder was not named on the policy, which facilitated the Design-Builder seeking recourse from the policy. In the current insurance market environment, and particularly when a prime designer is responsible to the project Owner for the development of documents to a specified level of completion prior to hand-over to the Design-Builder, project owners and underwriters generally prefer to have the Design-Builder and the design firms under contract to it also named on the policy. The insured versus insured exclusion on the policy precludes the ability of the Design-Builder to then seek direct recourse from the policy. This structure can leave the design activity performed by consultants under contract to specialty subconctractors unaddressed and potentially un- or underinsured.

TABLE 11.2   Dovetail standard policy exclusions to address interrelated exposures

Coverage	Exclusion	Treatment
A/E professional liability insurance	• Construction means and methods	• Delete or restrict to faulty construction workmanship
	• Equity interest	• Delete or set percentage of ownership threshold for coverage
	• Pollution	• Buy back or combine professional/pollution insurance
General liability insurance	• Professional errors and omissions (E&O)	• Carve out named perils or professional insurance
	• Pollution	• Buy back named perils or pollution insurance
	• Damage to insured's own work	• Negotiate coverage endorsement or cover under builder's risk insurance
Pollution liability insurance	• Professional E&O	• Combine professional and pollution insurance

Contractor professional liability wordings provide coverage for the acts, errors, or omissions of the Design-Builder in the provision of contractor professional services and grants coverage to the Design-Builder for damages arising from the acts, errors, or omissions of design professionals for whom the Design-Builder is legally liable. The Design-Builder (or the contractor party of a Design-Build team) will be the named insured.

The choice between these two approaches needs to be made in the context of the contractual regimes and delivery strategy: Is the Design-Build contract going to be executed by a team comprising a contractor and a design firm, or is the contract and risk allocation such that the Contractor fully accepts legal liability for the design (either through self-performance of design work in the case of an integrated Contractor or by means of subcontracts with design firms)?

If risk transfer is being sought via an A/E wording, careful consideration should be given to any exclusions for construction management and for construction means and methods. Increasingly, such wordings have narrowed coverage terms solely to "agency construction management services" for a fee and are of limited use as a vehicle for transferring construction management or program management professional services risk.

Contractors' pollution legal liability insurance coverage under an A/E wording is widely available but may be of limited utility on a subsurface project. It may have a professional services trigger, may be claims made, and will often not include nonowned disposal sites (NODS) and other transit hazards nor pre-existing pollution legal liability. Insurance rates for coverage via the professional liability market is also likely to be higher (and coverage terms narrower) than pricing and terms available from the pollution markets.

The majority of A/E professional liability policies on the market today also contain an exclusion for claims arising from pollution. As the demand for environmental design projects has grown, however, many of these insurers have become amenable to adding this coverage to their policies. Coverage for pollution exposure is typically provided through an endorsement to the standard professional liability policy that buys back the pollution coverage by eliminating or modifying the pollution exclusion.

The underwriting community has also introduced policy forms that combine a Contractor's pollution coverage with professional liability insurance. This approach is particularly well-suited for Design-Build environmental remediation projects. A key advantage to this approach is that a single policy covering professional and pollution liability exposures should eliminate the problem of disputes between insurers over which policy covers a particular loss.

Project-Specific Professional Liability. Both A/E and Contractor wordings are available on a project-specific basis. Underwriters consider Design-Build a problematic class of business to underwrite, particularly subsurface projects such as tunnels, which are subject to high rates-on-line and high retentions (typically $1 million to $10 million).

A/E-based wording will name the "entire design team" as the insured and is commonly used in a shotgun marriage scenario where teams are put together for one project without prior relationships. The Contractor and Owner will typically be recognized as an indemnified party, and joint defense provisions will apply. This is an expensive option because there is still a tendency for the Contractor member of the Design-Build team to seek to recover damages from the policy.

Contractor wording will name the Design-Builder and provide coverage via the insuring agreement for damages arising from the named insured's architects and engineers as well as the named insured's legal liability for design professionals under contract to them. Contractor wording will also better enable rectification provisions to allow the Design-Builder to address a potential design-based error during the course of construction in the absence of a formal demand.

Dedicated Project Insurance. For larger Design-Build projects, an increasing trend is for an Owner, Contractor, or the Design-Builder to procure dedicated project insurance (commonly known as *project-specific* or *wrap-up* insurance programs). A coordinated insurance program (CIP) provides a vehicle to protect the Owner, Design-Builder, and its subcontractors. Usual coverage provided under a CIP includes general liability, workers' compensation and employer's liability, excess liability, and builder's risk. Environmental liability and professional liability insurance may also be included and should be used for larger subsurface projects. A variation to this approach is a *rolling* wrap-up where several similar, medium to smaller projects can be combined under one blanket CIP.

The key benefits to a Design-Builder of a CIP include

- Coordinated insurance mechanism to address interrelated exposures,

- Reduction in the total cost of risk by instituting coordinated safety management and claims management activities, and

- Assurance that subcontractors/subconsultants have adequate limits and coverage in place.

Coordinated Insurance Programs. CIPs consolidate the general liability, workers' compensation, and builder's risk coverage for the Design-Builder and all its subcontractors/subconsultants into one program, insured by one carrier (per line) and managed by the Design-Builder. A number of risk management advantages are associated with CIPs:

- **Reduced insurance costs.** Because of the economies of scale, Design-Builders will enjoy more negotiating leverage with underwriters. Carriers can reduce the administration and service expenses with CIPs and, through competition, pass these savings on to the Design-Builder.

- **Expanded coverage and higher limits**. The same leverage-producing premium savings is used to broaden coverage. Favorable loss experience on projects protected by CIPs has underwriters eager to write these programs. This gives the Design-Builder significant leverage when negotiating coverage. For this reason, CIPs will provide higher limits and broader coverage to the Design-Builder and its subcontractors.

- **Centralized safety program**. Losses drive up project construction costs by increasing claim and litigation expenses, causing construction delays, and reducing efficiency. CIPs put the Design-Builder in a better position to influence project safety activities to reduce costs. By coordinating the safety program, the Design-Builder eliminates multiple insurance safety representatives with different levels of experience and dedication from the project. A CIP will improve the Design-Builder's position to negotiate better loss-control services from the carrier. With CIP, even the smallest subcontractors receive the attention usually reserved for the largest insureds.

- **Improved cash flow**. The CIP premium volume can result in favorable payment terms. With a high deductible or paid loss retrospective-rating plan, the Design-Builder may be able to hold significant amounts of insurance dollars until claims are actually paid.

- **Enhanced claims administration**. Design-Builders under CIPs will enjoy a higher standard of claim service. A CIP centralizes the claims administration process to ensure that claims are investigated properly and all claimants treated fairly.

- **Unified defense**. Projects insured under conventional programs tend to have a higher incidence of cross litigation. A significant amount of this litigation is initiated by insurance companies asserting subrogation claims on behalf of insureds. Because CIPs involve one underwriter, this reduces the carrier's ability and incentive to pursue such claims.

- **Improved minority business enterprise participation**. Many Owners or sponsors of large construction projects, especially in the public sector, encourage and/or require minority participation. Many of these smaller enterprises lack the required insurance coverages. A CIP diminishes this concern.

The use of CIP presents some challenges for the Design-Builder. For instance, some subcontractors may experience a diminished incentive to work safely because the bulk of the risk is transferred to the Design-Builder's program. This concern can be eliminated or reduced with an aggressive safety program. An incentive program that promotes the Design-Builder's risk management goals and objectives is preferred. Although there is concern that CIPs require a greater degree of involvement on behalf of the Design-Builder, a properly managed program will produce enough savings to justify the cost of the additional administration. Many benefits of CIPs are not quantifiable. Therefore, you should consider all the advantages of a CIP before deciding what method of risk management is best for the project. This is illustrated in Table 11.3.

How a CIP Works. The Design-Builder may procure the coverage on behalf of the subcontractors working on a project insured with a CIP. In return for this coverage, the subcontractors agree to remove all associated insurance costs, including profit, from the bids. Although the CIP provides workers' compensation, general liability, umbrella liability, and builder's risk insurance coverages, the Design-Builder controls the program and may expand or restrict the program as desired. For example, the insurance program may be expanded to also include

- Pollution liability,
- Railroad protective liability,
- Professional liability, and
- Owner's protective liability.

TABLE 11.3 Advantages and perceived disadvantages of CIP

Advantages	Perceived Disadvantages
Uniformity of coverage insurance	Assumption of safety
Economies of scale	Separate payouts
Savings through loss-sensitive programs	Duration of premium adjustments
Elimination of gaps, responsibilities, and overlays	Poor experience
Community relations, minority Contractor participation	Administrative burden
Elimination of coverage disputes	Loss of competitive edge
Reduced subrogation	Multiple audits
Improved safety program	Less incentive for safety
Effective claims and medical cost administration	
Centralized insurance	
Unification of interest	
Control of coverage	

Automobile liability insurance, Contractor's equipment insurance coverage, and surety bonds remain the responsibility of the individual subcontractors.

The financial incentives of CIPs are contingent upon the type of project, size, location, and legal environment. Most sponsors of CIPs enjoy direct savings of 20% to 40% over a traditionally insured program.

SURETY BONDS

Since the Design-Builder is the single-point authority, many Owners require some form of guarantee of performance and payment bonds. Parental guarantees and letters of credit are frequently used as well.

Performance Guarantees/Bonds

Performance bonds, one type of surety bond, are the most common form of security provided by Contractors working on domestic Design-Bid-Build projects. These bonds essentially provide that in the event the Contractor defaults in its obligations on the project, the surety will step in and fulfill the Contractor's obligations. The obligations may involve the surety procuring a new completion Contractor, paying for the Owner's procurement of a new completion Contractor, or financing and otherwise assisting with the original Contractor's completion of the work.

The use of performance bonds on domestic Design-Build projects mirrors (although to a lesser degree) their use on Design-Bid-Build projects. Performance bonds cover the full and faithful performance of all of the Design-Builder's obligations, including the design, unless specifically excluded by the bond.

Design-Build Surety Challenges

Most major surety writers will consider support of Design-Build contracts for their clients but will underwrite certain aspects of these contracts. The primary concerns of the sureties include:

- Underwriting a project that is yet to be designed, particularly if specific performance guarantees are associated with the work.

- Accepting an obligation for defective design risk. Sureties will want to determine the relationship between the designing party and the builder. Is it an in-house capability of the contractor? Will the Designer be a subcontractor? In any event, sureties will confirm professional liability insurance limits and terms of coverage to ensure balance sheet protection of the Contractor for any design liabilities that result from the contract.

- Addressing design performance/efficacy guarantees, which surety underwriters are not accustomed to and have difficulty quantifying the performance guarantee. Sureties will not bond long-term warranties (generally considered to be those exceeding five years) or certain efficiency warranties related to equipment or facility performance. Underwriters will questions the Contractor's insurance coverage for such exposures, as well as the adequacy of pass-through warranties to equipment suppliers and subcontractors. Further, sureties are unlikely to accept high liquidated damages and will not bond any damage provisions deemed to provide the Owner with actual damages protection.

- Determining the penal sum of the bond, given the fast-track nature of many Design-Build projects.

Larger Contractors with established surety relationships are in a better position to obtain performance bonds for their Design-Build projects, whereas engineering firms that lead the Design-Build team have a much greater challenge because they typically do not have established surety relationships. In addition, the capital base for these firms looks strikingly different compared to a traditional Contractor.

Environmental Surety

The surety market is even more restrictive for environmental Design-Build projects but surety support can be obtained for certain types of work. Given the changing conditions that inevitably arise during the remediation process, there are inherent uncertainties as to whether the design and remediation approach will meet the performance standards.

Approach to Surety Underwriters

When working with the underwriting community, there are several key issues, such as contract terms, qualifications of team members and financial conditions, to emphasize in order to enhance underwriters' comfort level with the Design-Build project.

Quality of risk transfer though insurance and contractual provisions: Because surety underwriters are concerned with the design efficacy exposure, it is important to demonstrate limits and scope of professional liability insurance maintained by the design professionals. Likewise, on environmental projects, underwriters will require that the Design-Builder and/or its subconsultants have in place pollution liability insurance (e.g., Contractor's pollution insurance or environmental E&O). Underwriters will also want to review the contractual obligations the Design-Builder has with the Owner as well as its subcontractors/subconsultants. The key contract terms to highlight are

- Scope of work;
- Pricing and payment;
- Change order process;
- Differing site conditions;
- Time and schedule management;
- Contractual risk allocation;
- Indemnifications performance guarantees/damages;

- Qualification and experience of the Engineer: the Design-Builder and its leading engineering firm, subcontractors' and subconsultants' level of experience in performing Design-Build project or similar type of work;

- Quality management programs that will mitigate the potential for a performance shortfall; and

- Financial standing of the project participants.

Subcontractor Default Insurance

The insurance program that enables the prime Contractor to manage the financial risk associated with the default in performance of subcontractors and suppliers is called *subcontractor default insurance* (SDI). Currently, there is only a single insurer source for this cover (Subguard, a coverage offered by Zurich). SDI is not appropriate for every contractor but can provide a cost-effective and efficient alternative to the traditional reliance on surety bonds from subcontractors and suppliers. SDI is an insurance policy, usually purchased by the prime Contractor, that directly indemnifies it for claims and damages resulting from the default in performance of a subcontractor or supplier. The SDI policy is designed as a large deductible insurance plan that provides the prime Contractor with security against catastrophic losses and with effective control over schedule, as well as the funding, claim, and recovery process. The SDI's main elements are

- Operates similar to a large-deductible liability program.
- Includes broad definition of covered losses:
  - Cost of completing work;
  - Payment of unpaid bills;
  - Cost of correcting deficient or nonconforming work;
  - Cost of delay damages and indirect costs;
  - Legal and professional costs associated with claim adjustment process; and
  - Indirect costs, subject to its own sublimit of coverage.
- Offers $50 million policy limit. Single subcontract coverage above $40 million requires pre-approval from the underwriter.
- Covers all subcontractors/suppliers that have executed a contract during the policy period.
- Offers a negotiable retention level, generally in the range of $500,000 to $1.5 million.
- Offers option of having the deductible applied per subcontractor or subcontract depending on the insured's appetite for risk.
- Makes loss aggregates available: annual or multi-year.
- Offers multi-year policy; generally coverage is written on an annual basis for a three-year term.

Most large prime Contractors manage the exposure of subcontractor or supplier defaults by a combination of pre-qualification and selection procedures and by transferring the risk to surety companies by requiring surety bonds from subcontractors and suppliers. The cost of transferring the risk of default to surety companies averages approximately 1% of the value of subcontractor/supplier contracts. This cost bubbles up into increased contract cost to the project, which means the total cost of subcontractor supplier bonds can be substantial. Unfortunately, the direct cost for surety bonds does not reflect the total cost of insuring the performance of subcontractors and suppliers and the financial consequences of nonperformance. Additional costs faced by the prime Contractor include the many hidden costs and inefficiencies that are inherent in the traditional subcontractor/supplier bonding process, including

- Inability of many subcontractors and suppliers to qualify for surety credit;
- Pressure on project profit margins, which prevents bonding 100% of subcontractors/suppliers;
- Delays caused by the surety's failure to respond promptly to claims;
- Delay and added cost of nursing "wounded" subcontractors before surety take-over;
- Friction caused by adversarial relationship with subcontractor's or supplier's surety, leading to disputes over the amount of damages recoverable;
- Uncollectable damages due to surety company default/insolvency; and
- Losses experienced by unbonded subcontractors.

FINANCIAL AND NONTRADITIONAL RISK COVERAGES

Insurance coverages for risks such as force majeure, liquidated damages, cost overruns, or other nontraditional financial risks are not currently available in the marketplace as they were prior to 2001.

BUILDER'S RISK INSURANCE

The purpose of builder's risk insurance is to cover the loss or damage to work in progress at the construction site. Because of varied insurable interests of the project participants, it is typical for one party to obtain builder's risk insurance for the benefit of all.

Some form of builder's risk coverage is usually required in Design-Build contracts. The insurance is intended to

- Cover all parties on the project,
- Apply on a replacement cost basis,
- Waive rights of subrogation (most insurance carriers will not waive subrogation rights against Architects and Engineers whose design error caused the loss),
- Apply until final payment is made, and
- Cover materials to be incorporated into the project while stored off-site or in transit.

It is important to note that while the term "all-risk builder's risk" is used in contracts, there are many exclusions that will vary from policy to policy.

Insurance companies may require adherence to *A Code of Practice for Risk Management of Tunnel Works*, prepared by the International Tunnelling Insurance Group (2006). The objective of the code is to promote and secure best practices for the minimization and management of risks associated with the design and construction of tunnels, caverns, and shafts. Intended to be a voluntary code, many carriers make it a provision under the policy. Noncompliance with the code would result in cancellation of insurance coverage, so it is important to determine if the project's insurance program is subject to the code's adherence.

Tunnel and Subsurface Construction Endorsement. One of the most important builder's risk insurance policy endorsements to review is the Tunnel and Subsurface Construction Endorsement. Sample wording follows.

When coverage is provided on tunnels and/or subsurface construction, the following additional exclusions apply:

- Expenses incurred for dewatering the insured project even if the quantities of water originally expected are substantially exceeded. However, if the peril of flood is insured under this policy, this exclusion shall not apply to water directly attributable to the peril of flood.

- Loss or damage due to breakdown of the dewatering system if the breakdown could have been avoided by means of adequate standby facilities.

- Expenses incurred for additional installations and facilities for the discharge of run-off and/or underground water.

- Expenses incurred for the repair of cracks in any concrete and leakage from the cracks, except that which results from a peril otherwise insured under this policy.

- Costs of grouting of soft mountain or rock areas and/or other additional safety measures, even if the need for such measures arises only during normal progress in construction.

- Costs of removing material that has been excavated in excess of the design and specifications and the additional resulting expenses for refilling of the voids and cavities.

- Expenses incurred for the repair of eroded slopes or other graded areas, if the insured has failed to take the preventative measures as required in the plans and specifications.

- Loss or damage due to subsidence if caused by insufficient compacting or grouting.

- Expenses incurred for modification in construction methods or technique, unforeseen ground conditions (differing site conditions), or obstructions.

- Loss or damage to tunnel boring machines, including abandonment or recovery of the machines.

- Loss of bentonite, suspensions, or any media or substance used for excavation support or as a ground-conditioning agent.

For subsurface projects, several important exclusion wordings should be reviewed. Several of these exclusions can and should be deleted; for example, the exclusion for loss or damage due to subsidence if caused by insufficient compacting or grouting, as well as the exclusion for expenses incurred for the repair of eroded slopes or other graded areas if the insured has failed to take preventative measures. Significant claims have arisen from subsidence and creation of voids above tunnels where insurance coverage was denied. Other uninsured claims have arisen from heavy rains eroding slopes of portal areas.

Coverage for loss or damage to tunnel boring machines can be included under the policy as well.

Financial consequences from delay in start-up or in completion due to the result of direct physical damage to the work can be endorsed to the builder's risk policy. This is usually a significantly high exposure in the case of tunnel structures due to a repair period that can exceed 12 months.

OTHER COVERAGES

Pollution Liability

Pollution liability insurance typically protects the insured against unanticipated losses associated with unknown pollution conditions, including cleanup costs and third-party property damage or bodily injury claims.

Policies can be designed to cover operational pollution risks arising from unanticipated discharges, leakages or spillages, and historical risks for liabilities associated with preexisting contamination. The cover can fill gaps arising from exclusions in general liability policies.

It is possible to combine both operational and historical pollution cover in a single policy. The policies can be extended to cover off-site waste disposal locations, transportation exposures, or even contingent risks such as business interruption or economic loss associated with contamination. When extensions of coverage for the transportation of contaminated soil were not specifically endorsed to the policy, uninsured claims have occurred.

Many policies also require specific locations of NODS in order to effect coverage.

Contractor's Pollution Liability

This is a form of pollution liability insurance designed to protect an organization facing environmental risks through their business operations—for example, environmental or construction contractors. Such operations may present an ongoing risk of causing pollution or exacerbating contamination conditions as a result of disturbing/remobilizing existing contaminants or following unanticipated discharges, leakages, spillages, and so forth.

Many project insurance specifications require specific pollution liability coverage. Contractors can arrange this on a portfolio basis or on a project-by-project basis. Coverage can be arranged on either a claims-made or occurrence basis.

Companies engaged in both environmental contracting and consulting operations often purchase contractor's pollution liability and professional liability/indemnity programs to efficiently package the pollution coverage for their professional exposure and to reduce the potential for gaps or overlaps in coverage.

Remediation Cost Cap

Cost-cap or stop-loss policies are designed to pay for unanticipated remediation project costs that exceed original project estimates. Cost overruns have many causes, including the discovery of additional contamination, underestimation of base costs, or changes in regulatory requirements. The insurance attachment point above which the policy will pay out (subject to the policy limit and conditions) is the subject of a negotiated agreement with the insurer.

Coverage is often written in conjunction with pollution legal liability insurance that may require a self-insured retention or copayments feature. A single market can write limits of $10 million or more per claim. This coverage has been useful to firms involved in brownfield remediation projects because it can cap the costs of remediation. This has provided protection to not only the firms conducting the cleanup but also the sellers and purchasers of the property and to the lenders providing funds to finance the purchase and cleanup.

Common Risks and Hazards

Common risks and hazards associated in construction and how they are transferred to insurance polices are presented in Tables 11.4 and 11.5.

TABLE 11.4   Insurance coverage summary matrix: common risks or hazards

Common Risks or Hazards in Construction Contracts	Who Is Exposed?	Method of Control
Loss of income due to death or injury to worker	Worker and dependent	Workers' compensation coverage—state and/or federal statutes (U.S. Longshore and Harbor Workers' Compensation Act [U.S. L&H] and Jones Act)
	Employer	Voluntary compensation and employer's liability for employees not required by workers' compensation law
Lawsuit brought by injured employee or dependent	Employer	Most workers' compensation acts eliminate the employee's right to sue employer within scope of act
		Employer's liability (coverage B or part two of standard workers' compensation policy would provide employer's liability coverage) or stop-gap insurance (employer's liability in monopolistic state-fund states)
Lawsuit brought by a member of the public or a party to a construction contract for damages arising from bodily injury, death, or damage to property of others and personal injury	Contractors, subcontractors, Owners, Architects, Engineers, and Construction Managers	• Commercial general liability insurance • Include coverage for liability assumed under contract • Professional liability insurance
Lawsuit brought by a member of the public or a party to a construction contract for damages arising from professional error or omission	Architects, Engineers, Construction Managers, consultants, attorneys, certified public accountants, and Contractors on Design-Build projects	Architects and engineers' professional liability E&O insurance
Lawsuit to recover damages for bodily injury or damages to property of others arising from use of an automobile owned by the defendant	Any auto owner or driver	Automobile liability insurance
Lawsuit to recover damages for bodily injury or damage to property of others arising from use of vehicle not owned by the defendant	Contractors, subcontractors, Owners, Architects, and Engineers	Nonowned automobile liability insurance
Low bidder unwilling to enter into a contract at the bid price and supply final bond(s)	Owner and Contractor	Bid bond
Contractor (or subcontractor) unable to turn over to Owner (or general Contractor) a project completed in accordance with the plans and specifications at the agreed price	*Directly:* Owner (or Contractor) *Indirectly:* commercial bank or mortgage company	Performance bond
Failure of contractor to pay accounts of those having a direct contract with it to supply materials or perform services	*Directly:* subcontractors, suppliers, and employees *Indirectly:* Owners, from public relations standpoint	Labor and materials payment bonds

Table continued next page

TABLE 11.4 Insurance coverage summary matrix: common risks or hazards (continued)

Common Risks or Hazards in Construction Contracts	Who Is Exposed?	Method of Control
Failure of Owner to pay Contractor	Contractor	Lien act or lawsuits
Construction delayed due to unforeseen circumstance	Owner and Contractor	Penalty-bonus clause or liquidated damages
Construction delayed due to a physical loss such as fire, flood, major item in transit, or special Contractor's equipment not readily replaceable, causing increased construction costs and extended financing charges during nonproductive period	Owner and Contractor (loss of profit or increased costs)	Consequential loss insurance or builder's risk soft-costs coverage
Accident frequency-type	Contractor	Contractor's equipment insurance: • Specified peril form • All-risk form • Auto physical damage insurance
Construction delayed due to strike, lockout, or nondelivery or nonperformance of key equipment	Owner and Contractor	Force majeure, efficacy, or systems performance relief events in contract clauses, limitation of liability
Construction losses due to fire, windstorm, theft, collapse, flood, earthquake, and so forth	Owner and Contractor	Builder's risk insurance with flood and earthquake coverages

RECOMMENDATIONS

Recommendation 11-1: To determine the true extent of coverage, it is important to review the standard policy exclusions and limitations.

Recommendation 11-2: Coverage for pollution exposure is typically provided through an endorsement to the standard professional liability policy that buys back the pollution coverage by eliminating or modifying the pollution exclusion.

Recommendation 11-3: Contractor wording will name the Design-Builder and provide coverage via the insuring agreement for damages arising from the named insured's architects and engineers as well as the named insured's legal liability for design professionals under contract to them.

Recommendation 11-4: Before deciding which method of risk management is best for the project, consider all the advantages of a CIP.

Recommendation 11-5: When working with the surety underwriting community, several key issues, such as contract terms, qualifications of team members, and financial conditions, should be emphasized in order to enhance underwriters' comfort level with the Design-Build project.

Recommendation 11-6: One of the most important builder's risk insurance policy endorsements to review is the Tunnel and Subsurface Construction Endorsement.

Recommendation 11-7: For subsurface projects, several important builder's risk insurance exclusion wordings should be reviewed. Several of these exclusions can and should be deleted; for example, for loss or damage due to subsidence if caused by insufficient compacting or grouting and for

TABLE 11.5   Insurance coverage summary matrix: special risks or hazards

Special Risks or Hazards in Construction Contracts	Who Is Exposed?	Method of Control

Use of aircraft	Contractor, subcontractor, and Owner	• Owned and nonowned coverage:
		  – Hull physical damage coverage
		  – Liability, including passenger liability
		• Waiver clause under Owner's insurance or Contractor asked to be named as additional insured on the aircraft owner's policy with respect to both hull and liability insurance
		• Accident insurance—death and disability
		• Cargo insurance (if not covered by all-risk builder's risk insurance)

Use of watercraft	Contractor, subcontractor, and Owner	• Delete watercraft exclusion from general liability policy • Marine insurance to cover hull, protection and indemnity, cargo
Marine construction projects	Contractor, subcontractor, and Owner	U.S.L&H and Jones Act coverage as required
Dangerous operations: • Blasting • Pile driving • Excavation (damage to subsurface utilities, etc.) • Underpinning, shoring, or removal of support • Demolition	Contractor, subcontractor, and Owner	Confirm no explosion, collapse, and underground exclusions exist under general liability and umbrella liability insurance policies or endorsements
Foreign operations	Contractor, subcontractor, and Owner	Varies with laws of country involved. A close check should be made with your insurance broker to determine type of workers' compensation coverage necessary and to determine that the liability and automobile insurance carrier is approved or admitted in the country involved. This is extremely important. Also, workers' compensation laws vary from country to country. In many instances, it is necessary to have two separate programs on workers' compensation insurance: one to cover international hires and one for stateside hires.

expenses incurred for the repair of eroded slopes or other graded areas if the insured has failed to take preventative measures.

Recommendation 11-8: *The pollution policies can be extended to cover off-site waste disposal locations, transportation exposures, or even contingent risks such as business interruption or economic loss associated with contamination.*

BIBLIOGRAPHY

Brierley, G., and Cavan, B. 1987. The risks associated with tunneling projects. *Tunneling Technol. Newslet.* 58. U.S. National Committee on Tunneling Technology, June.

International Tunnelling Insurance Group. 2006. *A Code of Practice for Risk Management of Tunnel Works.* January. http://www.munichre.com/publications/tunnel_code_of_practice_en.pdf. Accessed February 2010.

Loulakis, M., and Shean, O. 1996. Risk transference in design-build contracting. *Construction Briefings.* Federal Publications, April.

Mehr, R., and Hedges, B. 1974. *Risk Management: Concepts and Applications.* Irwin Publications.

Willis Construction Practice. 1996. *Cost of Risk Engineering.* Willis Corroon Publications.

CHAPTER 12

Summary of Recommendations

The practices recommended in this book for Design-Build subsurface projects are summarized here.

TEAM STRUCTURES AND RELATIONSHIPS

Recommendation 4-1: The Owner must clearly define the project and develop a detailed scope of services that it desires. Not having a clearly defined scope can lead to misunderstandings, disputes, and mistrust among the parties. In addition, an unclear scope will inevitably lead to scope creep, additional costs, and a disappointed Owner and Design-Build team.

Recommendation 4-2: The Design-Build team and the Owner must communicate during all phases of the project. The expectations of each party must be clearly communicated to the other.

Recommendation 4-3: Formal partnering should be adopted by the Owner and the Design-Build team as a way to handle the inevitable conflicts that will develop during the life of the Design-Build project.

Recommendation 4-4: The Design-Build team must have a teaming arrangement that clearly spells out the roles and responsibilities of each party, the deliverables and schedule for these, the relative financial participation and risks or rewards of each party, and how internal disputes will be resolved.

Recommendation 4-5: Each party to the Design-Build procedure needs to have a voice at the negotiation table so there is clear understanding of what is being asked and agreed upon and there are no surprises.

Recommendation 4-6: A joint Contractor/Designer risk assessment is an effective way to get both parties on the same side to avoid the "us versus them" trap.

Recommendation 4-7: A review by an outside insurance expert of the insurance products, project requirements, risk matrix, and teaming agreement is advised.

Recommendation 4-8: The Design-Build contract provisions should clearly delineate each party's risk responsibilities and include risk-sharing clauses (e.g., risk-sharing pools, provisional sums, and/or Contractor incentives for managing risk effectively). The party best able to manage or mitigate risks should be contractually responsible for them. In areas where risks are unknown or poorly defined at the time of bidding (such as unknown or poorly defined geotechnical conditions), provisions for price adjustment should be built into fixed-price–type contracts to help share cost-risk and provide equitable adjustment once such conditions or risks are better defined. This will minimize the need for unnecessarily large cost contingency in the bid price.

PROCURING THE DESIGN-BUILD TEAM

Recommendation 5-1: The Owner's team and its consultants need to clearly define all key project technical and management issues, thus reducing all uncertainties to as low a level as possible. At a minimum, the design should be carried to at least the 30% level, with adequate drawings, design criteria, and special specifications to define the finished product, while leaving means and methods largely to the Design-Build team. The geotechnical aspects should be nearly complete to save time and to reduce uncertainty.

Recommendation 5-2: The Owner must be familiar with formal procurement methods and the public and legal policies on which they are based before undertaking a large, complex procurement such as a Design-Build subsurface project. Careful selection procedures must include planning and a detailed and thorough documentation of the selection process, which must be effective and defensive to demonstrate the technical basis and objectivity in selecting the Design-Build team.

Recommendation 5-3: The Owner needs to understand the balance of control, time, cost, and technical approach that is required for a Design-Build procurement. Trust in the Design-Build team is essential to get the desired result.

Recommendation 5-4: The Owner should limit the number of qualified proposers because of the potential risk and financial expense involved.

Recommendation 5-5: The Owner needs to evaluate and decide, and make clear in the request for proposal (RFP), the selection methodology: best-value, lowest-price–technically-qualified, or low bid with basic qualifications.

Recommendation 5-6: The Owner must create a clear and concise but comprehensive RFP that defines all elements to be submitted. The Owner should be clear about what it needs, what risk it will retain, how the selection process will unfold, and how the project will be managed and controlled. A pre-qualification step that focuses on getting bidders that have experience both with Design-Build and of working effectively together in the past on similar projects is essential.

Recommendation 5-7: The Owner's team must communicate clearly to the proposers the evaluation criteria and how each is valued. Past performance on similar projects; design and construction expertise; legal, bonding, and insurance standards; and history of work-zone safety are all important and must be valued and weighed in a fair and reasonable manner. It cannot be understated how important it is to make these issues clear in the RFP.

Recommendation 5-8: As a general rule, a single contract package is preferable. Multiple contracts will create interfaces that are notoriously difficult to manage and that will increase management costs significantly.

Recommendation 5-9: To bring out the best teams and proposals, the Owner needs to evaluate the benefits and equity of a stipend for unsuccessful proposers. An honorarium will encourage more proposers after the pre-qualification stage and encourage a more detailed knowledge base for evaluation of the final proposals. An Owner should state whether or not all nonselected proposers will receive a stipend or just the top two or three.

Recommendation 5-10: A prudent Owner will create a specific set of Design-Build contract documents rather than modify ones used in Design-Bid-Build work. The Owner will have to supplement its in-house team with legal, technical, political, and economic specialists (experienced in

Design-Build projects) to adequately cover all aspects of risk definition and allocation, warranties, limits of liability, quality assurance/control enforcement, and so forth.

Recommendation 5-11: *The Owner must plan how to do the procurement early in the process. Objectives and constraints, especially in public projects, must be identified early and considered when planning an RFP for a subsurface Design-Build project. Legislation to allow the process may be necessary to go forward. Effective outreach to both the public and design and construction communities is also necessary.*

AGREEMENTS: OWNER–DESIGN-BUILDER AND DESIGN-BUILDER–ENGINEER

Recommendation 6-1: *Members of the Design-Build team should exchange concerns about proposed terms and conditions prior to finalization of a contract with an Owner and understand the risks posed by the contract, including but not limited to schedule and delay risks, onerous terms, and financial penalties, before final pricing occurs and commitments are made.*

Recommendation 6-2: *Once the project commences, it is critical that the subconsultants and subcontractors to the Design-Builder identify project issues as they occur and provide timely notice to the Design-Builder. In turn, the Design-Builder must take guidance from its team members (when appropriate) and notify the Owner when such issues and concerns are identified.*

Recommendation 6-3: *The agreements should contain an obvious and unambiguous memorialization of the allocation of risks, which should be clearly assigned through one of the project's contracts. The Owner should understand how the Design-Builder transfers and mitigates contractual risks through the services to be provided by other members of the Design-Build team.*

Recommendation 6-4: *Sureties and appropriate insurance should be used to backstop otherwise unacceptable risks. The Owner should also consider establishing contingency accounts to address risks that are beyond the control of project participants. In addition, the members of the Design-Build team offering professional services (i.e., the design engineers) should provide certificates of insurance providing evidence of sufficient coverage for the amount of reasonably foreseeable risk associated with a project.*

Recommendation 6-5: *Careful early planning of the entire project should be factored into developing the contracting architecture. Of particular importance in the early phase of the project is identifying and characterizing the interfaces between contracts. The Design-Builder must translate the entire scope of work negotiated with the Owner into clearly defined segments so that appropriate scope risk is assigned to the proper party and each party clearly understands its role in the entire process.*

Recommendation 6-6: *The generally accepted form agreements will likely require some modification to fit a subsurface project, and editing and modification should be accomplished by practitioners familiar with the underground construction industry.*

Recommendation 6-7: *The Owner and the Design-Builder should address (in the contract) an appropriate transfer of risk and responsibility for unique challenges associated with a specific project. To a certain extent, reasonable contingencies can be developed for issues such as material shortages, permitting delays, seasonal interruptions, and accommodations due to public needs.*

Recommendation 6-8: For the resolution of disputes, an effective agreement should include a fair and reasonable process that allows the parties to raise and attempt to resolve their issues and disputes without affecting ongoing project performance and completion.

DESIGN DEVELOPMENT

Recommendation 7-1: In order to minimize bidding contingencies resulting from uncertain scope and to avoid potential liability, the Owner should establish in as detailed a manner as possible the project scope in the tender documents, while still leaving the Contractor with the flexibility to develop innovative design solutions. In particular, the Owner should identify specific details needed to comply with existing third-party agreements and perform sufficient pre-tender geotechnical investigations to enable bidders to evaluate feasible means and methods of construction.

Recommendation 7-2: The Owner should recognize that the Design-Builder's level of effort required to complete the final design is not significantly different from that on a Design-Bid-Build project and should allow sufficient time in the project schedule to complete the design process.

Recommendation 7-3: In order to facilitate communication and identify conflicts at an early stage, the entire Design-Build team should go through a formal partnering program in which the goals and objectives of each party are set forth and recognized by the other participants.

Recommendation 7-4: The Owner is advised to take contract packaging considerations into account when determining delivery methods.

Recommendation 7-5: The Contractor and Designer that form the Design-Build entity must define the risk allocation between themselves and clearly identify when one is solely responsible for specific outcomes.

Recommendation 7-6: To increase the efficiency of the design process, the Owner should refrain from specifying detailed quality assurance processes in the RFP document but should consider allowing the Design-Build team to establish its own quality assurance plan and include it as a part of the technical proposal.

SUBSURFACE EXPLORATIONS

Recommendation 8-1: The Owner's geotechnical consultant should perform a phased subsurface investigation program during design. This program should address the needs of both the Owner's Engineer and the Design-Build team.

Recommendation 8-2: All geotechnical data obtained should be made available to shortlisted Design-Build teams prior to tender development.

Recommendation 8-3: Design-Build teams should retain their own geotechnical consultants during the procurement phase to assess the adequacy of the existing geotechnical database provided by the Owner, given the Design-Builder's planned means and methods. Design-Build teams should be allowed to request additional geotechnical investigations during the procurement phase as necessary to finalize their pricing for the work.

Recommendation 8-4: *The Owner's geotechnical consultant should implement supplemental subsurface investigations during the procurement phase and should provide that information to all tenderers before finalization of their proposals.*

Recommendation 8-5: *It is expected that the selected Design-Build team will need to perform its own geotechnical investigation after notice-to-proceed. This program should be subject to the oversight and acceptance of the Owner's geotechnical consultant, and the results of this program should be incorporated into the contract documents.*

GEOTECHNICAL REPORTS

Recommendation 9-1: *A thorough investigation of the subsurface ground and groundwater conditions should be carried out in advance of the selection of a Design-Build team.*

Recommendation 9-2: *Existing underground and overhead utilities should be addressed, including efforts by the Design-Build team to identify and relocate utilities in advance of construction.*

Recommendation 9-3: *All information learned and obtained in the procurement documents should be properly disclosed.*

Recommendation 9-4: *Design-Build teams should be engaged in a bilateral process during the bid phase that achieves a jointly prepared and agreed-upon Geotechnical Baseline Report (GBR) for construction.*

CONSTRUCTION PHASE ISSUES

Recommendation 10-1: *For a typical Design-Build contract, the usual principle that the Contractor takes responsibility only for those risks over which it has some control still applies. Specifically for subsurface work, "The Owner owns the ground" and differing site conditions clauses apply. The GBR will set out the conditions to be expected and the less likely but still possible conditions that the Contractor may have to deal with and for which it is compensated according to the contract.*

Recommendation 10-2: *During the construction phase, the risk register should be maintained as an active document, regularly updated to identify new risks and reassess existing ones, and to actively manage risks through the construction and commissioning phases and successful project completion.*

Recommendation 10-3: *When disputes occur, effective formal and facilitated partnering can lead the parties to resolve the dispute in a constructive manner and minimize the impact on progress of the work. The Design-Build contract should include an alternative dispute resolution provision for disputes that cannot be resolved between the parties, with the aim of speedy resolution through a fair and transparent process that all parties respect and are likely to accept.*

Recommendation 10-4: *To avoid lengthy design submittal reviews, rework by the Design-Builder's Designer, and disputes and delay claims, design workshops should be conducted in which the Owner's Consultants explain the development of the reference design and share their knowledge of the ground conditions, while responding to the Designer's proposed approach to the design and design innovations. Initial design workshops should cover the main areas of design, and additional workshops should be instigated by the Owner's Engineering Consultant when review of design submittals shows differences in approach or apparent misunderstandings of requirements.*

Recommendation 10-5: The Design-Builder's responsibility for quality control should be accompanied by responsibility for quality control—overseen by appropriate quality assurance audits. For underground construction, it is in the Owner's best interests to have its own forces or consultants perform some level of direct inspection and have an independent laboratory perform verification testing. The quality assurance audits should include direct observation of the inspection process and independent review of the data collected.

INSURANCE COVERAGE ISSUES

Recommendation 11-1: To determine the true extent of coverage, it is important to review the standard policy exclusions and limitations.

Recommendation 11-2: Coverage for pollution exposure is typically provided through an endorsement to the standard professional liability policy that buys back the pollution coverage by eliminating or modifying the pollution exclusion.

Recommendation 11-3: Contractor wording will name the Design-Builder and provide coverage via the insuring agreement for damages arising from the named insured's architects and engineers as well as the named insured's legal liability for design professionals under contract to them.

Recommendation 11-4: Before deciding which method of risk management is best for the project, all the advantages of a coordinated insurance program should be considered.

Recommendation 11-5: When working with the surety underwriting community, several key issues, such as contract terms, qualifications of team members, and financial conditions, should be emphasized in order to enhance underwriters' comfort level with the Design-Build project.

Recommendation 11-6: One of the most important builder's risk insurance policy endorsements to review is the Tunnel and Subsurface Construction Endorsement.

Recommendation 11-7: For subsurface projects, several important builder's risk insurance exclusion wordings should be reviewed. Several of these exclusions can and should be deleted; for example, for loss or damage due to subsidence if caused by insufficient compacting or grouting and for expenses incurred for the repair of eroded slopes or other graded areas if the insured has failed to take preventative measures.

Recommendation 11-8: The pollution policies can be extended to cover off-site waste disposal locations, transportation exposures, or even contingent risks such as business interruption or economic loss associated with contamination.

Some Thoughts on Design-Build for Subsurface Projects

This appendix was submitted by the staff of Parsons, based in Pasadena, California.

INTRODUCTION

Design-Build for tunnels and underground projects is clearly here to stay as an alternative to the traditional Design-Bid-Build. There are certainly other alternative contract delivery systems, such as construction management at risk, general contractor construction management, early contractor involvement, and cost-reimbursable contracts, which have their own advantages and disadvantages whether you are the Contractor, Owner, or Designer.

The primary Design-Build drivers and catch words often cited by Owners seeking to go Design-Build in the underground environment are litigation control, new technologies, Contractor innovation, risk management, privatization, cost savings, schedule acceleration, strategic alliances, politics, Designer and Contractor in close alliance, and providing a single-point contract. Design-Build is cited as avoiding the pitfalls of over-conservatism in the Design-Bid-Build setting as the team works together to reduce risk of claims and produce a more constructible and economic design. The negative sides cited by Owners are higher cost of bidding, higher profits for the Contractor and Designer, less control over the final product, a perceived more difficult procurement process, less quality, and concern of adequate bidders.

Historically, Design-Build has worked successfully for large complex projects—dams, cathedrals, opera houses, and industrial projects. For the underground, Design-Build certainly has its risk due to the nature of tunnels; mostly linear construction and risks of unpredictable ground means one disruption can shut down the whole operation. Some lessons learned from Parsons' various experiences as Owner's representative, joint venture partner, Designer, bidder, and independent Engineer on the Design-Build contract delivery method on small and large subsurface Design-Build projects follow.

NORTH FORK STANISLAUS HYDROELECTRIC PROJECT FOR CALAVERAS COUNTY WATER DISTRICT AND NORTHERN CALIFORNIA POWER AGENCY

This project was in planning and design since 1957. In 1984, the Federal Energy Regulatory Commission (FERC) license was issued as Design-Bid-Build, and in 1986 was tendered as Design-Build and completed in 1989. The project consisted of over 10 miles of tunnel, a 280-ft high rock-fill dam, and 185-ft high combined gravity arch dam, which was the first in 30 years to be permitted by the State of California Division of Safety of Dams.

Lessons Learned

- Owner was a small agency without the infrastructure to manage and monitor the project, so Design-Build was ideal contract delivery mechanism, particularly after 27 years of planning.

- Bidding Contractors needed to pool money to do a few basic boreholes to check for fatal flaws. However, excessive over-excavation at the gravity arch dam led to cost overruns. The bid had line items for potential changes, but no differing site conditions (DSC) clause, and the winning contractor, Sierra Constructors (Electrowatt-Guy F. Atkinson), was the optimistic bid.

- FERC requirements required Tudor Engineering Company–CT Main (Parsons) quality assurance/control (QA/QC) on top of the Contractor's program, so the constructed project had excellent quality. Moreover, with dam safety always an issue, there was understandable concern about over-conservatism in design. The right amount of QA/QC is a recurring theme in Design-Build. Refer to ASCE Civil Engineering (May 1990) for more about the project.

CENTRAL LINK FOR SEATTLE SOUND TRANSIT

In 2000 King County's Sound Transit (Washington state) tendered a twin tube subway Design-Build project to link the downtown (Seattle) bus tunnel with the University of Washington. The subsurface geotechnical conditions for the project, considered a high risk, employed a Geotechnical Characterization Report that was well done and left little to the imagination (unlike international Design-Bill tenders), and a Tender Geotechnical Baseline Report (TGBR). Attempts by contractors to negotiate a final GBR were unsuccessful. Only two Contractor teams bid, and the winning bid was Modern Transit Constructors (Modern Continental-SA Healy/Impregilo-Dumez with Parsons/D2Consult/GZA as Designers).

It was about a quarter billion over the half billion dollar engineer's estimate, and value engineering was implemented to see if the costs could come down. A mono tube with a splitter wall and cavern stations vs. twin bore with binocular stations was proposed, which resulted in approximately a 15% savings. Even with this savings, the costs were above the project budget. The Owner's budget was over 2 years old without escalation, did not reflect the significant concessions given to stakeholders, and did not consider the lack of an agreed upon final GBR (which required contingencies in the Contractor's bid). The client's management errors were documented by state auditor reports, and misrepresentation of budget was a setback for the Design-Build tunnel. The project was shelved and rebid in sections largely as a conventional Design-Bid-Build in 2009.

Lessons Learned

- An accurate Owner's budget and estimates are important no matter what the contract delivery mechanism.

- The level to which ground conditions are baselined should not be so broad as to never permit any DSC changes. Doing so defeats the point of having a GBR because the ground risk remains totally with the Contractor.

RANCHO BERNARDO PIPELINE 6

This was one of more than 90 capital improvement projects in the City of San Diego Water Utilities Capital Improvement Program, which the City of San Diego elected to experiment using the Design-Build process. The project included 50,000 ft of 2-to-4-ft diameter reclaimed and potable water pipelines and one conventional tunneled 400 ft-long creek crossing. The Contractor was Black Mountain Road Pipeline Contractors (Archer Western/MDEC). The city is a big agency with an extensive management, design, and operations and maintenance infrastructure, but it nevertheless wanted to try the Design-Build procurement for the often-cited Design-Build advantages. As program manager, Parsons and the city identified their misgivings.

Lessons Learned

- Design-Builder should show demonstrated experience where Designer and Contractor have worked successfully together, including a contractual agreement, which allows the Designer to be fully integrated and heavily involved at every phase of construction, not just shop-drawing review.

- If control is desired, Owner should have language in contract requiring submittal of all design changes in a timely manner to allow adequate review time.

- Preparation of request for proposal (RFP) should clearly define the design criteria and "quality" of project desired in the RFP and referenced program standards and guidelines.

- For QA/QC, Design-Builder's role should be defined in detail, such as for weld and compaction testing by the desired party specifically identified as being responsible.

- The city's expectations needed to be defined: that is, a shortened schedule and contained costs while allowing creative design solutions.

Refer to ASCE *Civil Engineering* (February 2003) for more about the project.

TRANSPORTATION EXPANSION PROJECT (T-REX)

A $1.7 billion highway and light rail expansion of I-25/225 in Denver was owned by the Colorado Department of Transportation (CDOT) and the Regional Transportation District (RTD) and constructed by Southeast Corridor Constructors (Kiewit Parsons, JV) with additional oversight by the Federal Highway Administration, Federal Transit Administration, and the City and County of Denver. Design-Build was chosen for fast track and the agencies' ability to manage the mega-project. One element of the project was the rerouting of many storm drains, including a newly tunneled New Mississippi Avenue Outfall, which was critical to avoid short- and long-term flooding of the highway. The 30% RFP design identified a longer outfall, performance tunneling specifications in difficult ground conditions crossing obliquely to the freeway and under costly right-of-way (ROW) properties.

Agencies' documents allowed for an alternative configuration concept (ACC), and methods not used before by CDOT or RTD would require special permission. At bid time the ACC was not pushed forward, mainly due to low cover issues at one of the bridge crossings. The Design-Builder, however, continued to push after award. The original ACC provided for earth pressure balance (EPB) tunneling using a one-pass segmented liner and a complete revised shorter alignment. The revised ACC used a multi-structure approach, requiring a combination of shallow EPB pipe jacking, EPB tunneling, shallow cut and coverbox, and underpinning of three railroads.

The proposed changes allowed for the best solution, saved millions on tunneling, avoiding costly ROWs (and headaches), and employed a tunneling method never used in Colorado and a tunnel liner with features never before used in North America. The job was not without problems, however, but these would have happened in any contract setting. Tunneling by nature is Design-Build, but the contract offered no DSC, unless Type 2 DSCs were encountered. Some Type 1 DSCs were encountered in open-cut work at the outlet but were small compared to the benefits.

Lessons Learned

There was a true partnership by all parties. ACC provisions and a sophisticated Owner allowed for innovations but not without problems. Refer to *Engineering News-Record* cover story (September 15, 2003) and ASCE *Civil Engineering* (September 2003) for more about this project.

LAKE HODGES TO OLIVENHAIN RESERVOIR PIPELINE

The $35 million, 6,000-by-10-ft diameter power tunnel is part of an $800 million emergency storage capital improvement program for the San Diego County Water Authority, a medium-sized agency with a well developed management and operations infrastructure. The program is under an OCIP and PLA. Some items were taken to 90% design and some to 30% design or less, averaging 60% complete design, when the decision was made to go Design-Build. The drivers for Design-Build were to implement a demonstration project, realize potential schedule savings, and leave vertical alignment decisions and whether the tunnel be driven by TBM or drill-and-blast to the Contractor. The contract incorporated a GBR for bidding and required a GBR for construction, with an allowance for $1 million if no DSC were requested. The winning contractor, Kiewit in association with Parsons, chose a conventionally driven tunnel with slopes up to 19.6%, completed on schedule, and filed no DSC issues.

Lessons Learned

- A medium-sized agency can implement Design-Build.
- The contract structure for geotechnical issues was an incentive and allocated risks fairly.
- Letting the marketplace decide design issues can be done in Design-Build and Design-Bid-Build settings, but tunneling by its nature lends itself to Design-Build with the Contractor taking ownership for design decisions.

CONCLUSIONS

Other subsurface Design-Build projects whose lessons are yet to be learned at Parsons vary from the small $23 million Sandy River Conduit Relocation Project (Portland, Oregon), which is nearly complete; the $447 million intake Pump Station No. 3 (Southern Nevada Water Authority); and the mega $8.7 billion Access to the Region's Core Trans-Hudson Express Tunnel projects (New Jersey Transit). The industry will also have lessons learned on the upcoming public-private partnership Port of Miami Tunnel and the Alaskan Way Viaduct tunnel in Washington and other future mega-transportation and water projects. The industry is getting better at it, and, certainly, Design-Build seems appropriate for mega-projects, no matter the size of the Owner's organization, for a variety of reasons cited in the introduction.

The higher risk for tunnel projects (whether Design-Build or Design-Bid-Build) is that they differ from other projects because construction is linear, which can have a domino effect if something goes wrong. Also, specialized Contractors are relatively few compared to other building industries. Moreover, the tunnel alignment is often predetermined by other project requirements, whereas the geology of alignment has been determined by nature. The chances of DSC can be very high. Although the U.S. National Committee for Tunneling Technology (1984) recommends 3% of costs for urban tunneling be spent on geotechnical investigations to help reduce unknowns, sampling is less than 1/10,000 of the ground that will be encountered.

Sophisticated procurement, quality-based selection (% dollar/% quality), innovation/ value engineering/partnering/risk management with unit cost allowances, fair allocation of geotechnical risks, high level of QA/QC, and Owner sophistication can be achieved and provisions specified in either and any contract delivery mechanism. There is no free lunch with Design-Build; you just exchange one set of issues for another. Design-Build is not any more sophisticated or necessarily better than other contract methods. It is just different and therefore requires some upfront homework by all parties.

Finally, Design-Build solicitation packages are crucial elements in the process, and the percentage complete should be clearly defined (typically, a minimum 10% to 30%) with site plan surveys, specific engineering design criteria, general and special contract requirements, and fair allocation of geotechnical risks (considering some of the lessons learned and previously cited). They should include tools to manage risks during construction, such as partnering, dispute review board, GBR, etc.; that is, employing the best contractual practices recommended by A. Matthews over 35 years ago and the revised recommendations by the Underground Construction Association of SME (2006).

Also, there are some particular advantages to requiring separate project manager/construction manager review and administration in the solicitation package, defining upfront project oversight by Owner or other stakeholder to ensure expectations and harmony during construction, and being able to manage and facilitate rapid and effective decisions.

In the end, our work is about the contract, communication, and developing trust and true partnership, regardless of the contract delivery mechanism.

The tunneling industry is necessary for our civilization's sustainability, and we are very lucky to be able to have this discussion and be a part of it.

Gerald A. Bonner, Ph.D., P.E.C. Eng.
Senior Program Director
Technical Director of Tunnels

Chris W. Dixon
Vice President
Construction Manager

Kevin W. Ulrey
Construction Director
SNWA IPS No. 3

Jerry H. Ostberg, P.E.
Senior Construction Manager

Lukas B. Wendel, P.E.
Principal Project Manager

Fred B. Estep, P.E.
Principal Construction Manager

William R. Phillips
Vice President
Manager of Construction Support Services

Donald S. Miner
Retired Vice President
Technical Director for Tunnels

Jon Y. Kaneshiro
Senior Engineering Manager
Technology Leader for Tunnels

Mike M. Feroz, P.E.
Construction Director

Roger Rothenburger, P.E.
Construction Manager
SNWA IPS No. 3

Shimi Tzobery, P.E.
Construction Manager

David R. Yankovich, P.E., S.E.
Principal Project Manager

David L. Rendini
Senior Construction Manager

Philip B. Colton
Senior Construction Manager

Stephen J. Navin, P.E.
Retired Director of Underground Management

Index

Date Due

FEB 29 2012			
DEC 1 4 2013			